ESSENTIALS OF
E-COMMERCE TECHNOLOGY

V. RAJARAMAN
Supercomputer Education and Research Centre
Indian Institute of Science
Bangalore

PHI Learning *Private Limited*
New Delhi-110001
2011

₹ 250.00

ESSENTIALS OF E-COMMERCE TECHNOLOGY
V. Rajaraman

© 2010 by PHI Learning Private Limited, New Delhi. All rights reserved. No part of this book may be reproduced in any form, by mimeograph or any other means, without permission in writing from the publisher.

ISBN-978-81-203-3937-8

The export rights of this book are vested solely with the publisher.

Second Printing ... ... **July, 2011**

Published by Asoke K. Ghosh, PHI Learning Private Limited, M-97, Connaught Circus, New Delhi-110001 and Printed by Mudrak, 30-A, Patparganj, Delhi-110091.

In fond memory of
Srimati S. Rajyalakshmi
Internet Savvy at 86
Akka to everyone
Full of zest for life

Contents

Preface xi

1 WHAT IS ELECTRONIC COMMERCE? 1–15
    1.1   Introduction 1
    1.2   Types of E-Commerce 3
            1.2.1   Business to Customer E-Commerce 3
            1.2.2   Customer to Customer E-Commerce 6
            1.2.3   Business to Business E-Commerce 6
    1.3   Advantages and Disadvantages of E-commerce 9
    1.4   Supply Chain Management in E-commerce 12
    Summary 12
    Exercises 13
    Objective Questions 13

2 INFRASTRUCTURE FOR E-COMMERCE 16–38
    2.1   Introduction 16
    2.2   Local Area Network 19
            2.2.1   Interconnecting LAN Segments 22
    2.3   Public Switched Telephone Network 24
            2.3.1   Broadband Connection to Home PC 25
            2.3.2   ISDN Service 26

	2.4	Cable Network	26
	2.5	Wireless Networks	27
	2.6	Microwave and Satellite Network	28
		2.6.1 Satellite Communication	29
	2.7	Private Communication Networks	31
	Summary		32
	Exercises		34
	Objective Questions		35

3. COMMUNICATION NETWORKS FOR E-COMMERCE 39–56

- 3.1 Introduction 39
- 3.2 The Internet 40
- 3.3 Packet Switching and the Internet Protocol 40
- 3.4 Transmission Control Protocol 43
- 3.5 Domain Names 43
- 3.6 Intranet and Extranet 45
  - 3.6.1 Intranet 45
  - 3.6.2 Extranet 46
- 3.7 Firewalls 47
- 3.8 The Future of Internet Technology 49
- *Summary* 50
- *Exercises* 51
- *Objective Questions* 52

4. NETWORK SERVICES 57–82

- 4.1 Introduction 57
- 4.2 The World Wide Web 58
  - 4.2.1 Hypertext 58
  - 4.2.2 Universal Resource Locator 60
- 4.3 Information Retrieval from the World Wide Web 62
  - 4.3.1 Search Engine Based Retrieval 62
- 4.4 Markup Languages 63
  - 4.4.1 Hypertext Markup Language (HTML) 64
  - 4.4.2 Web Pages with Tables 66
- 4.5 Other Network Services 68
  - 4.5.1 File Transfer 69
  - 4.5.2 Telnet 69
  - 4.5.3 E-Mail 70
- 4.6 Advanced Web Technologies 72
  - 4.6.1 Multiple Windows in Browser 72
  - 4.6.2 Common Gateway Interface Program 73
  - 4.6.3 Cookies 73
  - 4.6.4 Active Documents 74
  - 4.6.5 Cascading Style Sheets 74

	4.7	Conclusions	75
		Summary	75
		Exercises	77
		Objective Questions	78

5. SECURE MESSAGING 83–115

- 5.1 Introduction 83
- 5.2 Symmetric Data Encryption with Private Key 84
  - 5.2.1 Digital Encryption Standard 87
  - 5.2.2 Triple DES Encryption 90
  - 5.2.3 Advanced Encryption Standard 91
- 5.3 Public Key Encryption 92
  - 5.3.1 RSA Encryption Scheme 93
  - 5.3.2 Diffie–Hellman Key Exchange Algorithm 96
  - 5.3.3 Combining RSA with DES 98
- 5.4 Public Key Certifying Authority 99
- 5.5 Digital Signature 101
- 5.6 Some Applications of Encryption Technologies 103
  - 5.6.1 Secure E-mail 103
  - 5.6.2 Secure Socket Layer 104
  - 5.6.3 Secure Hypertext Transfer Protocol 104
- 5.7 Conclusions 105

Summary 105
Exercises 107
Objective Questions 109

6. PAYMENT SYSTEMS IN E-COMMERCE 116–150

- 6.1 Introduction 116
- 6.2 Requirements of e-payment Systems 117
- 6.3 Credit Card Payment 118
  - 6.3.1 Credit Card Payment using Secure Socket Layer 120
  - 6.3.2 Secure Electronic Transaction (SET) Protocol 121
  - 6.3.3 Dual Signature Scheme 122
- 6.4 Electronic Funds Transfer 124
  - 6.4.1 Automated Cheque Clearance 125
  - 6.4.2 Electronic Clearing Service 126
- 6.5 Electronic Cheque Payment 127
  - 6.5.1 Electronic Clearance of Pay Order 129
  - 6.5.2 E-cheque Format 131
- 6.6 Electronic Cash 131
  - 6.6.1 E-Cash Issue and Spending 132
  - 6.6.2 Anonymous E-cash 134
  - 6.6.3 Smart Card-Based Cash Payment 135

viii Contents

6.7	Payment Gateways	137
	6.7.1 Pay Pal	137
6.8	Micro-payments for Information Goods	138
6.9	Conclusions	139
Summary		140
Exercises		141
Objective Questions		143

7. STRUCTURED ELECTRONIC DOCUMENTS 151–172

7.1	Introduction	151
7.2	Electronic Data Interchange	152
	7.2.1 Transporting EDI Formatted Data	155
7.3	EDI and XML	156
7.4	XML and HTML	160
	7.4.1 XSL Style Sheet	161
7.5	Basics of Web site Design	162
Summary		165
Exercises		166
Objective Questions		167

8. M-COMMERCE 173–205

8.1	Introduction	173
8.2	Layered Architecture for m-commerce	174
	8.2.1 Mobile Phone—SMS System	175
	8.2.2 Laptops using Wi-Fi LAN Systems	175
	8.2.3 WAP-Enabled Mobile Hand-held Systems	176
	8.2.4 Location Dependent Services	176
8.3	Mobile Communication Infrastructure	177
	8.3.1 Architecture of GSM Cellular Mobile Wireless System	177
	8.3.2 General Packet Radio Service (GPRS)	180
	8.3.3 CDMA 1xEVDO Rev.A	180
	8.3.4 Short Message Service (SMS)	181
8.4	Wireless Application Protocol	181
	8.4.1 Mobile Network Operators	182
	8.4.2 Mobile Handset Manufacturers	182
	8.4.3 Service Provider	182
8.5	WAP Gateway	183
	8.5.1 WAP and i-Mode	185
8.6	Wireless Markup Language	185
	8.6.1 XHTML	187
8.7	Secure Wireless Connectivity	188
	8.7.1 Security of Mobile Network-Internet Connection	188
	8.7.2 WAP Gateway Managed by Sensitive Content Providers	189
	8.7.3 WAP Gateway at Server End	189

    8.8   Mobile Payment Methods 190
            8.8.1   SIM Card-enabled Payments 190
            8.8.2   Payments based on SMS 190
            8.8.3   Payment using WAP-enabled Mobile Hand-held Device 191
    8.9   Mobile Banking 193
    8.10 Conclusions 195
    Summary 196
    Exercises 197
    Objective Questions 199

9   E-COMMERCE OF MULTIMEDIA 206–225

    9.1   Introduction 206
    9.2   E-Publishing of Multimedia 208
    9.3   Digitizing and Storing of Books 209
    9.4   Digitizing and Storing Audio 210
    9.5   Digitizing and Storing Video 210
    9.6   Distribution of e-books 211
    9.7   Distribution of Audio 212
    9.8   Video on Demand 213
    9.9   Intellectual Property Issues 215
    Summary 219
    Exercises 219
    Objective Questions 221

10 LEGAL FRAMEWORK OF E-COMMERCE 226–236

    10.1 Information Technology Act 2000 226
    10.2 Information Technology (Amendment) Act 2008 229
    Summary 232
    Exercises 233
    Objective Questions 234

References *237–238*
Answers to Objective Questions *239–242*
Index *243–246*

Preface

A number of developments in computers and communications have led to the rapid growth of the Internet. Among several applications of the Internet, e-commerce has emerged as the most important one. E-commerce has grown rapidly in India. Booking e-tickets for travel by trains and airlines has now become a routine procedure.

The basic objective of this book is to briefly describe the technologies which have made e-commerce a success. There are several books on e-commerce and one might wonder the need for yet another book. Most of the available books are aimed at students of schools of management and do not discuss technology in depth. Besides, they are verbose and descriptive. My aim is to briefly present the essentials of the subject with emphasis on technology. The book will give a quick introduction to a beginner covering all important topics.

The book consists of ten chapters. It begins with a chapter which answers the question: What is electronic commerce? In this chapter, B2C, B2B and C2C e-commerce are defined with several examples. It is followed by a chapter which describes the infrastructure needed for e-commerce. It presents a layered architecture of e-commerce systems with six layers. Each layer can be independently designed, but depends on the services provided by the layers below it. This chapter describes the physical layer to provide connectivity such as local area networks, wireless networks and communication satellites.

The third chapter is on communication networks which is primarily the Internet. The following chapter on Network Services describes the World Wide Web and services provided by it. It specifically introduces html and the basics of web page design. Chapter 5 is on secure messaging. This chapter describes various symmetric cryptographic algorithms including DES, Triple DES, Advanced Encryption Standard. It also discusses RSA and

Diffie–Hellman public key algorithms. Digital signature which is essential to authenticate e-documents is also described. Finally, the public key certification method is described.

The sixth chapter is devoted to various methods of payment for goods and services in e-commerce. In this chapter, protocols used for credit card payment including the Secure Electronic Transaction (SET) protocol which uses the dual digital signature scheme, are discussed. E-cash payment methods including the technology available to make anonymous e-cash payments are explained.

The seventh chapter is on structured electronic documents which can be exchanged using either the Internet or intranet between business partners in B2B e-commerce. It starts with a discussion on Electronic Data Interchange (EDI) standard and enumerates its drawbacks. It then describes the use of XML as an alternative to EDI. The chapter is concluded with some tips to design a good home page for organizations.

The growth of mobile hand-held devices in India has been phenomenal. Thus, a whole chapter is devoted to explain m-commerce. It presents a layered architecture for m-commerce which is somewhat different from that for e-commerce. Of special interest is the detailed description of Wireless Application Protocol (WAP) which is used in mobile Internet. One major concern in m-commerce is security. This chapter discusses the appropriate technology to provide security. It also describes various mobile payment methods.

The ninth chapter is normally not found in most e-commerce books. This is the emerging market for multimedia e-commerce. The distinct feature of this mode of e-commerce is that it delivers the items purchased electronically using either fixed or mobile communication infrastructure. This chapter describes the recently introduced mobile e-book delivery method introduced by Amazon using Kindle, an e-book reader. It also describes Video on Demand systems which are now appearing in India.

The last chapter is on the legal framework essential for e-commerce. India is one of the few countries which enacted the Information Technology Act in 2000 to provide the legal framework. The act provides for a public key certifying authority and defines various cyber crimes. Based on experience with implementing the Act, the Government of India has amended the Act in 2008. This chapter critically examines the amended Act and its implications in e-commerce.

The book is designed to be used as a textbook by students of BCA, MCA, B.Sc./M.Sc. (Computer science and information technology), BE courses in information technology, and computer science and engineering. It has several student-friendly features. Each chapter has learning goals and a summary. Besides around 300 exercises, there are nearly 350 objective type questions. The book will also be useful to IT professionals for a quick introduction to the subject. The overall structure of the book is designed to promote self-study. The chapters are written in an easy-to-understand style. Of special interest to the students are the objective type questions which are framed to provide an enhanced understanding of the subject. An answer key to the objective questions is provided at the end of the book.

A book of this type has gained insight from several books and journal articles.. There is a set of references given at the end of the book which lists the books referred by the author.

I would like to thank several of my colleagues at IISc, Bangalore who critically reviewed the material sent to them and provided valuable feedback which has been incorporated. Prof. Y. Narahari gave valuable comments on the whole book. Prof. C.E. Veni Madhavan provided valuable feedback on Chapters 5 and 6 relating to security and payment systems.

Mr. H. Krishnamurthy meticulously read the whole book and gave several valuable suggestions. Mr. Chakshu Ray of Parliamentary Research Services, New Delhi, critically read Chapter 10 on legal framework of e-commerce and provided valuable feedback. Sincere thanks are due to Ms. T. Mallika for patiently word processing several drafts of this book and for secretarial assistance. Finally, I would like to express my affectionate appreciation to my wife Dharma for critically reading the book, editing it and for lending dedicated support for writing this book.

V. RAJARAMAN

CHAPTER 1

What is Electronic Commerce?

LEARNING GOALS

In this chapter, we will learn:
1. How e-commerce evolved.
2. Classification of e-commerce as Business to Customer (B2C), Business to Business (B2B) and Customer to Customer (C2C).
3. Advantages and disadvantages of e-commerce.

1.1 INTRODUCTION

During the past five years, a number of advances in technology have combined to lead to novel applications of computers. The speed of processors has continued to increase. Gordon Moore who predicted in the 1970s that the speed of computers will double every 18 months has been proved right. Now (in 2009) Intel has announced a Pentium processor with a clock speed of 3.4 GHz! Simultaneously the speed and capacities of main Random Access Memories have also been increasing. The capacity has been doubling every eighteen months and today a desktop PC has at least 512 MB memory. The secondary memory size has also been increasing, doubling almost every 12 months. Today a PC has a minimum of 80 GB disk. All these together have made an individual's desktop computer a really powerful machine. The increase in speed and capacity has in turn made it possible to write some truly innovative and powerful software which has made it easy to use the PC. One should specifically mention the graphical user interfaces. The toolbar, icons and the mouse together has made it possible for users to invoke many applications by pointing to icons using a mouse and clicking. The keyboard seems to be becoming a secondary device! While the hardware and software improvements of individual computers have been going on, concurrently better hardware and software have emerged to interconnect computers. The value of networking computers was realized as it allows sharing resources efficiently. It was observed by Metcalfe that the utility of an individual computer increases as the square of the number of computers to which it is connected. In other words, if 8 computers are interconnected by a local area network, the utility of an individual computer will increase by a factor of 64. This law is reasonably simple to understand from the analoging of mobile phones. The utility of your mobile phone increases rapidly as more people are connected to you and you can talk to any one of them. The first good standard for interconnecting computers within an organization (which is still being used) is the Ethernet standard which has stood the test of time. The next step after the emergence of local area network was the Internet, a worldwide interconnection of local networks. The Internet idea germinated in the late 1960s but really became feasible as a large-scale system only after the emergence of desktop PCs. The Internet use exploded after the development of World Wide Web (which is an application run on the Internet) and browsers used to access resources available in the web.

Lastly, speed of communication between computers increased rapidly even without replacing the basic communication infrastructure. Communication speeds are doubling every nine months with the cost remaining same. For example, a modem which connects a PC to the telephone line provided by BSNL had a speed of 1200 bits per second (bps) in the mid-1980s; today it is 256 Kbps without any change of telephone lines. This is achieved by efficient use of telephone lines and providing more processing power to modems which allows better data encoding and error detection/correction. This speed is expected to increase further to megabits per second range. Another development is WiFi (Wireless high fidelity) which allows mobile laptop computers to connect wirelessly to the Internet. The use of mobile phones in India is growing by over 50% each year for the past six years and they can be connected to the Internet using General Packet Radio Service or by using Wireless Application Protocol (WAP).

The three advances, that of processors, user interfaces and faster communications led to many innovative applications which used the Internet infrastructure. Of these, electronic commerce (abbreviated as e-commerce) is probably the most significant as it will change the way we do business.

Definition of e-commerce

We may briefly define electronic commerce (e-commerce) as: "sharing of business information, maintaining business relationships and conducting business transactions by using computers interconnected by a telecommunication system". The reason we have used this definition will become clear at the end of this chapter. World Trade Organization (WTO) has defined e-commerce as "the production, distribution, marketing, sale or delivery of goods and services by electronic means". Our definition is more general. We assume in our definition "business transactions" to include the production, marketing, distribution, sale or delivery of goods and services.

1.2 TYPES OF E-COMMERCE

The telecommunication system used in e-commerce is the Internet which connects computers all over the world which can communicate with one another using well-defined rules (called a *standard protocol*).

The Internet by itself remained the preserve of academic institutions for several years until its commercial potential was understood. This led to the emergence of a number of Internet Service Providers (called ISPs) who provide value-added services to their customers using the Internet infrastructure. The general public was thus able to use the Internet and the number of Internet users increased by leaps and bounds all over the world particularly with improved communication infrastructure. This spawned a number of innovations in business between commercial organizations, between individuals and commercial organizations and between individuals. These transactions are commonly known as Business to Business (B2B), Business to Customer (B2C) and Customer to Customer (C2C). Another new entrant is Government to Customer (G2C) and Government to Business (G2B) in which individuals and business organizations can transact business with the government using the Internet. Business transactions include orders sent to vendors to supply items, invoices sent by vendors, payments made by credit cards, payments through electronic funds transfer and cash payments made using what is known as electronic cash. The important point is that all transactions are carried out electronically using the network infrastructure. There are a variety of e-commerce applications. Some of these are as follows:

- Retail sales of goods such as books, CDs, toys, etc.
- Auction sites using which an individual buyer and seller can buy and sell goods.
- Railways, airlines, hotels, etc., which permit booking on-line and payment by credit card.
- Banks providing information using the Internet on the status of accounts to customers.
- Cooperating businesses carrying out transactions in a semi-automated manner using the Internet.
- Filing tax returns with government agencies and obtaining immediate acknowledgement.
- Web-based education by delivering lessons on-line, on-line examinations, and certification.
- Delivering music, video entertainment (movies, sports events, etc.) and books via the Internet or an appropriate communication medium for all those subscribing for the service.

Every day new ideas seem to be emerging in electronic commerce which addresses its wide reach and round the clock functioning.

1.2.1 Business to Customer E-Commerce

When an individual buys items from a shop using the Internet and the entire transaction is carried out electronically, we call it business to customer e-commerce and it is abbreviated as B2C e-commerce. The shops which transact business using the Internet are called by various names, some of which are e-shop, virtual store, dot com shop and cyber shop. One of the earliest e-shops was a bookstore called amazon.com set up in

USA which primarily sells books and has now added other items such as gifts, music CDs, etc. Currently many similar shops are being established in India of which indiabook.com is an example and sells books and CDs. All e-shops have the following common characteristics.

1. Persons who want to shop have to use the Internet. They may have Internet access from home or workplace or an Internet kiosk (i.e., public Internet access point such as cyber cafes). A customer normally knows the web address of the shop with whom he or she wants to transact business and typically uses a web browser such as Netscape or Internet Explorer and enters the web address of the shop.

2. The home page of the shop is displayed which provides various options to a customer. If a customer wants to buy a book, he or she keys in its particulars. He or she may also request books available on a particular subject in which case the shop would search its database and give a list of books available on that subject. The shop may also display the contents page of a book selected by the customer, reviews of the book, its cost and discount if any.

3. If the customer wants to buy one or more books, he or she points to the book details displayed on the screen using the mouse and clicks. The vendor's computer enters the prices of the book(s) selected by the customer, provides discount (if any) and displays the net amount payable. The customer enters the shipping address and payment is usually by credit card. The credit card number is entered which is used for charging the customer. Sometimes option is also given to pay cash on delivery.

   Credit card payments are more common. As the Internet is not very secure, it is necessary to hide the details of the credit card number from snoopers and also from the merchant. The credit card details should be available only to the bank authorizing payment. It is done by using a protocol called Secure Electronic Transaction protocol (abbreviated as SET protocol) which we will describe later in this book.

4. The credit card details entered by a customer is sent in an encrypted (i.e., secret coded) form over the Internet and is forwarded to the authorizing bank by the merchant.

5. If the credit is OK, the bank authorizes the transaction.

6. The e-shop acknowledges the order and gives the details of delivery period and mode of shipment as desired by the customer.

7. The e-shop may not stock the items in its warehouse. It sends an electronic request to the distributor to ship the items either directly to the customer or to the e-shop for packing and forwarding. If it is a fast moving item, e-shop will normally stock the item and despatch it to the customer.

8. The credit card company's bank credits the shop's bank account electronically and sends a bill to the customer.

In Figure 1.1, we show the various steps in B2C e-commerce.

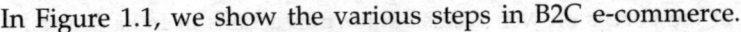

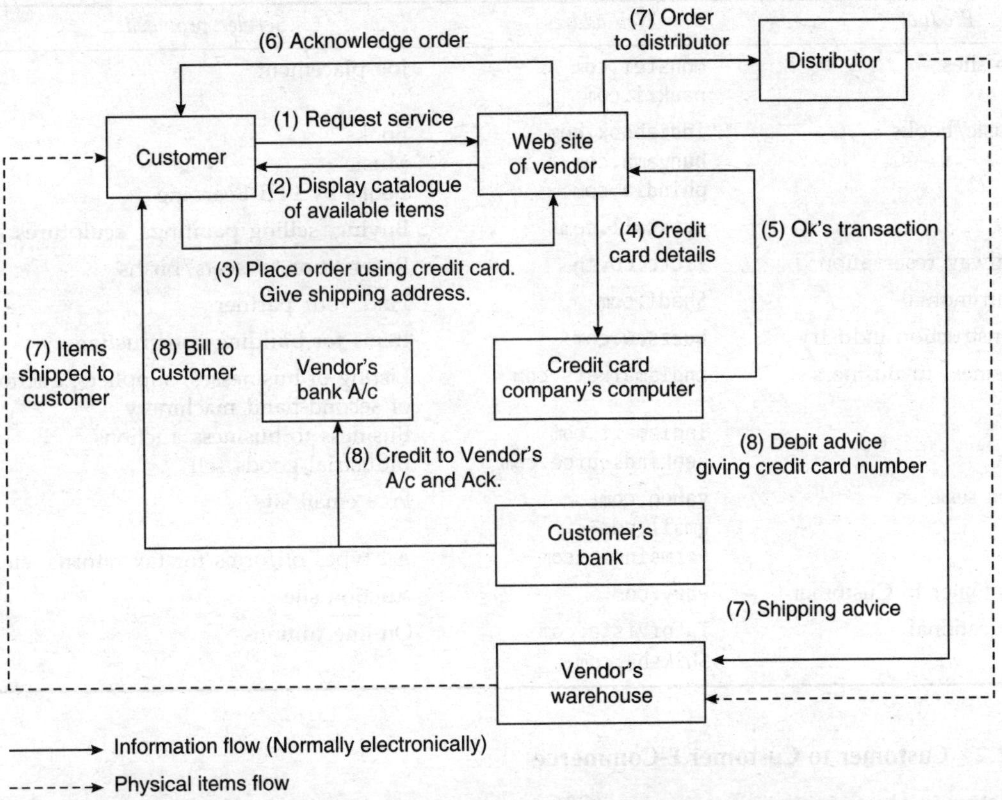

Figure 1.1 Block diagram of business to customer e-commerce.

There are a large number of sites or portals on the World Wide Web in India for various goods and services. We list some of them in Table 1.1.

Table 1.1 Some of the portals on web accessible for transactions in India

Product	Web address	Service provided
Automobiles	indiacar.com	New and used car sales
	automartindia.com	Used cars
Finance	indiainfoline.com	Investment advice
	walletwatch.com	Personal finance
	moneycontrol.com	Personal finance
	equitytrade.com	Investment advice
General shopping	rediff.com	Over 15 categories of goods
	indiashop.com	Toys, miscellaneous items
	beautyarcade.com	Cosmetics
	fabmall.com	General goods

(Contd.)

Table 1.1 Some of the portals on web accessible for transactions in India *(Contd.)*

Product	Web address	Service provided
Job sites	monster.com naukri.com	Job placement
Music/books	indiabook.com hungama.com phindia.com	Books Music Books by PHI Learning
Art	artstall.com	Buying, selling paintings, sculptures
Railway reservation	irctc.co.in	Reservation of seats/births
Matrimonial	Shadi.com	Find your partner
Construction industry	buzzsaw.com	Items for building construction
Business to Business	indiamarkets.com	Listing of businesses, suppliers, auctions of second-hand machinery
	indiamart.com	Business to business auctions
	seekandsource.com	Industrial goods sell
Free services	yahoo.com gmail.com	Free email site
	formsindia.com	All types of forms for tax returns, etc.
Customer to Customer	ebay.com	Auction site
Educational	TutorVista.com Shiksha.com	On-line tuitions

1.2.2 Customer to Customer E-Commerce

Customer to customer e-commerce (C2C e-commerce) is one in which two individuals want to sell/buy items. The items are usually used items, collector's items such as stamps, coins or antiques. The seller posts the description of the item and the expected price of the item on a web site maintained by a company which acts as a broker. An individual who logs on to this site looking for items may be interested in the item advertised for sale. He/she then offers to buy the item and may quote a price. The price is mutually settled between the two parties by exchanging messages through e-mail. The broker then arranges to collect the item from the seller and despatches it to the buyer and collects payment for the item and a fee from the buyer and the seller for services. The primary advantage of this transaction is that the Internet enables two individuals located at distant places to come together to buy and sell using an intermediary's web address.

1.2.3 Business to Business E-Commerce

Business to business e-commerce (B2B e-commerce) is perhaps the most important of the three e-commerce modes. It is growing very fast and it is predicted that most businesses in the world will participate in B2B e-commerce during this decade. We will illustrate B2B e-commerce with an example of a business purchasing goods electronically from a vendor. The two parties are the purchaser and the vendor. Businesses normally have their own local area network which connects all computers of their organization. A purchase transaction initiated by a purchaser proceeds as follows:

1. A purchase order is entered by the purchaser's office using a desktop computer and transmitted by e-mail to the vendor. A standard format for purchase orders sent by e-mail which is called Electronic Data Interchange (EDI) standard may be used or a mutually agreed format may be used. Such a format for a purchase order may also be created using a language called XML (Extended Markup Language). There is also a need to sign the purchase order electronically to meet legal requirements. We will describe EDI standard, XML and digital signature later in this book.
2. When the purchase order is received, the vendor immediately acknowledges it electronically. Observe that the purchase order need not be entered manually again by the vendor's clerk. This is in contrast to current computerized systems in which a vendor's clerk has to enter manually the purchase order on a PC (when it is received) for further processing. Manual entry is slow, prone to errors and expensive as a clerk's time has to be allocated for this. The electronic order also called a *soft copy* can be interpreted by the vendor's computer using a program provided a standard mutually agreed format is used by both the purchaser and the vendor. The inventory database is now searched for the availability of ordered items and appropriate action is taken to (electronically) acknowledge the purchase order. The inventory database of the vendor is updated and a delivery note is prepared to be sent to the receiving office of the purchaser. Concurrently an invoice for items supplied is transmitted by e-mail to the purchaser's accounts office. There is also a standard format for electronic invoices. Such a format is required by the purchaser's computer program to interpret the invoice and process payment. The vendor despatches the items physically along with a hard copy of delivery note. The delivery note is needed for physical inspection of items received and reconciling them with electronic delivery note and invoice to facilitate payment.
3. The items received from the vendor are sent to an inspection office of the purchaser along with the delivery note. The inspection office physically checks the items for both quantity and quality and sends a discrepancy note of items rejected to the purchase office.
4. The accepted items are sent to the store along with an electronic intimation. The stores office takes items into stock and also updates the inventory database. Simultaneously, the purchase office is intimated to enable it to handle rejected items and to authorize accounts departments to pay for the accepted items.
5. The accounts office electronically pays for items accepted and taken into stock. Electronic payment is made by the accounts office by informing its banker to debit authorized amount from its account and credit it to the vendor's bank account. The vendor's account may be in a different bank. The electronic transfer of funds from one bank account to another bank account is an important aspect of e-commerce. This is called Electronic Funds Transfer (EFT) and is already used in India for dividend payments, interest payment, etc. Observe that if EFT is used, no hard copy of cheques are sent by mail and the funds transfer is fast.

The above procedure is illustrated in Figure 1.2. This example also illustrates the need for good computer and communication infrastructure both within an organization and for interconnecting organizations.

8 *Essentials of E-Commerce Technology*

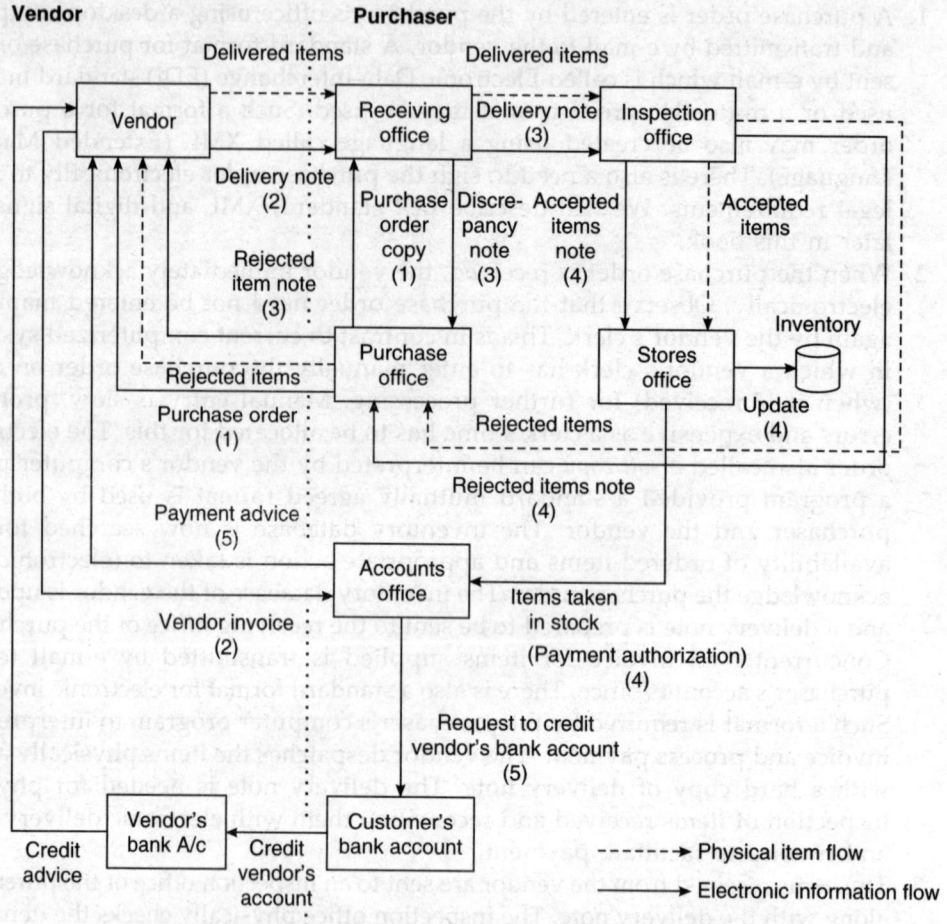

Figure 1.2 Block diagram of document, item flow and funds transfer in B2B e-commerce.

The following are the essential requirements for B2B e-commerce.

1. All businesses must have a LAN interconnecting computers in their respective offices and the offices themselves should have computers for data entry/receipt, comparison, etc. The system may be a distributed client/server type system with each office being a client and the databases being stored in appropriate servers. The internal system architecture is not a major issue. However, the protocol used by the LAN is usually the same as that used by the Internet, namely, TCP/IP. The organizational computer network using this protocol is called an *intranet*. Besides using TCP/IP protocol, intranets also have one or more World Wide Web servers. A business must already have a good operational computer-based information system for it to successfully participate in B2B e-commerce.
2. The two intranets must be interconnected. There are three alternative methods available to do this. One is to connect each of the intranets to the Internet which is an inexpensive solution but may be insecure. The second solution is to connect the intranets using a private communication network to constitute what is known

as an *extranet*. Extranet is a private interconnection of the intranets of business associates. Each intranet supports web pages which can be accessed by the members of the extranet. The third alternative is to interconnect the intranets of the two businesses by what is known as a Virtual Private Network (VPN). A VPN uses the Internet but implements special security measures to protect data. We will describe VPNs in greater detail later in this book.

3. There must be an agreement on a method of paying for goods or services received electronically. This implies that the business partners must know one another's bank account details. Further, funds transfer must be secure and each transaction should be authorized electronically.
4. We have assumed that documents are interchanged by e-mail. This is acceptable between close business associates but is not secure and there is no authentication of documents sent and received. For more secure transactions e-commerce has introduced data encryption and a method of digitally signing documents which we will discuss later in this book.

1.3 ADVANTAGES AND DISADVANTAGES OF E-COMMERCE

We saw in the last section that a large number of goods are now being bought and sold using the World Wide Web. If B2C is to grow fast, we require a good Internet connectivity to homes. This is currently poor in smaller towns of India. There are, however, many advantages of B2C and C2C e-commerce which are enumerated below:

1. One can buy/sell items from anywhere in the world using one's computer and Internet connection. Transactions can go on 24 hours a day, 7 days a week as the servers maintained by businesses to cater to e-commerce are usually never switched off.
2. Besides goods, services such as financial, legal and medical consultation may also be obtained using the World Wide Web infrastructure. Anonymous friendly advice may also be available on items one would like to buy.
3. A wide variety of goods, particularly items such as books and music, are accessible from e-shops. Besides saving the effort and time to visit a shop, an e-shop provides many services which are not usually provided in a physical shop. For example, an electronic book shop would provide a list of books you selected, display a list of similar books (i.e., books on the same topic), comparative review, sample chapters, readers' comments, etc. You can use this information before you decide to buy a book. The number of books available in such a shop is almost limitless as there is no physical shelf space and display needed. It is also possible for an e-bookshop to observe the pattern of buying of books by regular customers and alert them when similar titles are published. There is also a trend to publish only electronic versions of books without any hard copy as it is less expensive and faster to publish and distribute. E-books are sold and sent using a mobile network nowadays in e-form which can be read using a specially designed e-book reader.

One of the most popular services in B2C commerce in India is booking of train tickets comfortably from your home avoiding waiting in long lines at the railway booking office. The web site irctc.co.in maintained by the Indian Railways permits you to book tickets on any train from any place to any place in India. Payment is made by credit card and the ticket is delivered at your doorstep by courier or an e-ticket may be printed by a customer. It is estimated that currently the monthly ticket sale is Rs. 8 crores and every day 3000 tickets are booked on-line. Nowadays all domestic airlines encourage e-booking of tickets and most air tickets are e-tickets.

The major advantages which accrue to businesses that participate in e-commerce are as follows:

1. Businesses can reach out to customers worldwide at low cost. A well-designed web page will be an asset to any business to publicize their goods and services and also to sell their merchandise.
2. Order processing time and cost are reduced as manual entry of data is reduced. When a vendor receives a purchase order, he or she need not reenter it on his system for data processing. Businesses are also carried out faster as electronic exchange of documents is instantaneous across the world.
3. A manufacturing organization requires components which are supplied by several vendors. For example, a car manufacturer will receive supplies of tyres, electrical items, nuts, bolts, etc., from several vendors. There are thousands of such items and tracking and regulating supplies is a complex task. This is called Supply Chain Management (SCM). SCM becomes simple and fast in e-commerce. When cooperating businesses are connected by an extranet or VPN, regular suppliers can observe the production plans and inventories of a manufacturer and adjust their supply schedules, inventories, etc. This is possible as remote log in and access to databases of cooperating businesses may be mutually agreed. This reduces inventory sizes leading to cost reduction of goods.
4. Electronic funds transfer is fast and safe.
5. A large number of potential business partners can be quickly found and contacted by searching the World Wide Web.
6. Certain types of goods can be customized and sold directly by a manufacturer or assembler eliminating middlemen. For example, Dell Computers sells PCs directly to customers configuring them as per individual's requirements. Middlemen such as distributors and retailers are eliminated. Supply chain management is improved as manufacturers can adjust their inventory level of items and order processing based on customer orders in hand and customer preferences gathered over a period of time.
7. The cost of setting up an e-commerce site is quite small compared to the cost of having large premises.
8. The cost of transactions is quite low. It is estimated that transaction costs are less than a fifth of that of traditional business.
9. There are some items such as airline tickets where competing airlines provide several special packages and prices. Quick comparison is possible on the World Wide Web and a confirmed booking obtained on-line. For airlines also it is an easy method of attracting a large clientele without expensive promotional expense.
10. Companies can maintain on-line e-catalogue of items and a price list which can be quickly updated. This is much cheaper than printing and distributing catalogues and price lists. Electronic versions can be changed quickly when new items are added/deleted.

11. Customers can be serviced on-line expeditiously.
12. There are industries such as automobile manufacturers where a large number of ancillary units produce subsystems such as horns, brake linings, tyres, lights, etc. E-shopping to get the best deals and delivery has now become very common in B2B commerce. Intermediaries also provide auction sites which allow expeditious price negotiation. This allows lowering of costs and so called just-in-time manufacturing.

The major disadvantages of e-commerce are given as follows:
1. Currently Internet access is not widely available in India. Communication infrastructure is expensive and not very reliable particularly to individuals. This is, however, improving.
2. Payment by credit card requires faith in the system security. Customers are wary of giving their credit card numbers to vendors who have only a "web presence". Secure credit card transactions in which credit card numbers are encrypted and sent to a vendor are essential.
3. Electronic Data Interchange standards have to be in place before business to business e-commerce can increase. Small businesses may find it difficult to conform. Data interchange using XML (a new document description language) is expected to solve this problem.
4. Many persons go shopping for social contacts, touch and feel and bargaining before buying items. E-commerce will de-personalize transactions.
5. A major concern is security of transactions on the Internet. Spies or hackers can steal and misuse credit card numbers, purchase orders, invoices, etc., if appropriate care is not taken.
6. Shopping portals will be vulnerable to attacks by hackers unless special precautions are taken. One type of nuisance is called denial of service in which a large number of frivolous enquiries are posted to a portal making it inaccessible to legitimate customers.
7. Portals have to be protected by special security systems from virus attacks and other electronic vandalism and espionage.
8. Customers' privacy may be lost if regular log is kept of their buying habits.
9. The web site of vendors should have the capability of being scaled up quickly when the number of users suddenly increases. If the server's capability is limited, the response time of the site will be unacceptably high if a large number of customers decide to use the site. Thus, a vandor should be able to add more servers quickly when this happens.
10. If there is a sudden increase in orders, there may be logistical problems in physically delivering items to customers. Long delays in receiving ordered goods will adversely affect future sales. Being prepared to handle seasonal surge in demand requires pre-planning.
11. When a successful e-business is launched, immediately there will be many copy cats who will attempt to duplicate it. Duplication is much simpler in e-commerce compared to traditional business as it is easy to quickly build a web site and start a competitive business. Thus, to maintain the advantage of being first with the idea requires continuous innovation and improvement.

12. On-line businesses expose their catalogues and price lists to competitors. The advantage of secrecy of traditional mode of doing business is lost.
13. Not every item is suitable for sale in the web. For example saris, fancy furniture, etc., which require touch and feel are unsuitable for sale through the web.

In spite of these disadvantages, e-commerce is bound to rapidly increase due to its convenience, cost saving and wide reach.

1.4 SUPPLY CHAIN MANAGEMENT IN E-COMMERCE

We have already mentioned briefly the use of e-commerce in supply chain management. As this is one area of operation where e-commerce has made a significant difference, we describe it in greater detail in this section.

A good supply chain management ensures having the right product, at the right time, at the right place and at the right price. The primary aim is to reduce the volume of unsold items and ensure that items are not out-of-stock when required. Most organizations participating in e-commerce would like to have guaranteed delivery of items when required. There are two strategies that can be followed. If it is a commonly available commodity, then one can follow an e-auction path. In this strategy a request for e-quotation is publicized in the web on the organization's web site or on an intermediary site specializing in B2B auctions. Based on the quotes the best option can be picked. This can be done periodically. Based on sales forecasts, delivery schedules can be arranged to ensure just-in-time availability. This will minimize the cost of procurement and minimize inventory holding cost.

If the item is specialized where it is important to develop vendors, a cooperative strategy can be followed. In this case, the cooperating vendor can be allowed access to the demand forecasts and the actual stock position in the database via the extranet or VPN. Using this information the cooperating vendor can plan the production schedule to meet the expected demand. When the stock position of the purchaser goes down, the vendor knowing the position can automatically prepare to deliver the items and intimate the purchaser electronically. Such a system ensures for the purchaser and vendor a well-coordinated seller–buyer relationship and both parties gain. This is possible only with an extranet or VPN connection and trusted relationship.

SUMMARY

1. The sharing of business information, maintaining business relations and conducting business transactions by using telecommunication networks is defined as e-commerce.
2. E-commerce is normally categorized as Business to Business (B2B), Business to Customer (B2C) and Customer to Customer (C2C).
3. Development of e-commerce requires a good computing and communication infrastructure in a country.
4. The major advantages of e-commerce are: any time, anywhere transaction, reduction in cost of transactions, reduction in time to market a product, faster intra-organization communication and faster transfer of funds.

5. The major disadvantages of e-commerce are poor security of transactions, danger from hackers and reduction in human contacts.
6. Due to advantages being more than disadvantages, e-commerce will thrive in the coming years.

EXERCISES

1.1 Describe the advances in technology which has facilitated e-commerce.
1.2 Define e-commerce. What are the different types of e-commerce?
1.3 Explain B2B e-commerce using an example of a book distributor who stocks a large number of books and distributes them via a large number of booksellers. Booksellers order books as and when their stock is low. Distributors give one month's time to booksellers for payment.
1.4 Explain B2C e-commerce of a customer reserving airline tickets from home or workplace.
1.5 Explain C2C e-commerce with an appropriate example.
1.6 List the advantages and disadvantages of e-commerce.
1.7 What is supply chain management? How does e-commerce help in supply chain management?
1.8 What do you understand by e-governance? Give an example of G2C e-governance.
1.9 What infrastructure is needed if a business wants to start a B2B operation?
1.10 What infrastructure is needed at a customer's premises if he or she wants to use B2C e-commerce?

OBJECTIVE QUESTIONS

Each question has four possible answers. Pick the most appropriate answer.

1.1 By e-commerce, we mean
  (a) Commerce of electronic goods
  (b) Commerce which depends on electronics
  (c) Commerce which is based on the use of the Internet
  (d) Commerce which is based on transactions using computers connected by telecommunication network

1.2 For carrying out B2B e-commerce, identify the infrastructure required.
  (i) World Wide Web
  (ii) Corporate intranet
  (iii) Electronic Data Interchange Standards
  (iv) Secure Payment Services
  (v) Secure electronic communication link connecting businesses
  (a) (i), (ii), (iii)
  (b) (ii), (iii), (iv), (v)
  (c) (i), (ii), (iii), (iv), (v)
  (d) (ii), (iii), (iv)

1.3 For carrying out B2C e-commerce, identify the infrastructure required.
   (i) World Wide Web
   (ii) An intermediary portal
   (iii) Electronic Data Interchange Standards
   (iv) Secure Payment Services
   (v) Secure electronic communication link connecting businesses to customers
       (a) (i), (iv)   (b) (i), (iii), (iv)
       (c) (i), (ii), (iii), (iv)   (d) (ii), (iii)

1.4 For carrying out C2C e-commerce, identify the infrastructure required.
   (i) World Wide Web
   (ii) An intermediary portal
   (iii) Electronic Data Interchange Standards
   (iv) Secure Payment Services
   (v) Secure electronic communication link connecting customers
       (a) (i), (ii)   (b) (ii)
       (c) (i)   (d) (iii), (iv)

1.5 Advantages of B2C e-commerce to businesses are:
   (i) Business gets a wide reach to customers.
   (ii) Payment for services becomes easy.
   (iii) Shop can be open 24 hours a day seven days a week.
   (iv) Privacy of transaction can always be maintained.
       (a) (i), (ii)   (b) (i), (ii), (iii)
       (c) (i), (iii)   (d) (iii), (iv)

1.6 B2C e-commerce
   (a) Includes services such as legal advice and businesses selling products
   (b) Means only shopping for physical good
   (c) Means only customers should approach customers to sell
   (d) Means only customers should approach business to buy

1.7 Advantages of B2C e-commerce to customers are
   (i) A wide variety of goods can be accessed and comparative prices can be found.
   (ii) Shopping can be done at any time.
   (iii) Privacy of transactions can be ensured.
   (iv) Security of transactions can be ensured.
       (a) (i), (ii)   (b) (ii), (iii)
       (c) (iii), (iv)   (d) (i), (iv)

1.8 Disadvantages of e-commerce in India are
   (i) Internet access is not universally available.
   (ii) Credit card payment security is not yet fully understood.
   (iii) Transactions are de-personalized and human contact is missing.
   (iv) Cyber laws are not in place.
       (a) (i), (ii)   (b) (ii), (iii)
       (c) (i), (ii), (iii)   (d) (i), (ii), (iii), (iv)

1.9 By intranet, we mean
   (a) An extension of the Internet
   (b) An organizational network
   (c) An organizational network which uses TCP/IP protocol
   (d) A private organizational network which uses TCP/IP protocol

1.10 By an extranet, we mean
   (a) An extra secure intranet
   (b) The intranets of two organizations connected by the Internet
   (c) The intranets of several cooperating organizations interconnected by a secure private connection
   (d) The intranets of several cooperating organizations connected by Internet

CHAPTER 2

Infrastructure for E-Commerce

LEARNING GOALS

In this chapter, we will learn:

1. A methodology of examining e-commerce systems as a layered architecture.
2. About the various parts of the lowest layer in the e-commerce architecture, namely, the physical layer.
3. How a Local Area Network (LAN) works.
4. How a Public Switched Telephone Network (PSTN) is used in e-commerce.
5. About the use of cable networks in e-commerce.
6. About the use of wireless systems such as cellular networks, microwave stations and Very Small Aperture Satellite Systems in e-commerce.

2.1 INTRODUCTION

When one examines a complex system, it is a good idea to break it up into a number of parts where each part has a specific function to perform. These parts may be arranged as a number of layers. This is similar to the way one may divide a building into several layers each having a function. The bottommost layer is the foundation—a very important part; over this layer one has a floor. Walls are built over the foundation and enclose the floor. The walls, in turn, support the ceiling. Each layer has a function and provides support to the upper layers. E-commerce systems may also be thought of as consisting of many layers, each layer providing a service. Each layer has a specific function and can be described separately. This gives us a logical way to discuss the architecture of e-commerce systems. One possible layered architecture is given in Table 2.1. We have used six layers to logically discuss e-commerce systems. Each layer has a function and supports layers above it. The bottommost layer is the physical layer. By a physical layer, we mean the physical infrastructure such as cables, wires, satellites and mobile phone system. Their common property is that they provide the communication infrastructure for e-commerce. In fact, without high speed, reliable electronic communication e-commerce is not possible. The emergence of wireless communications has enabled one to use mobile hand-held phones and computers which has resulted in the emergence of mobile commerce abbreviated as m-commerce. We will describe the physical layer in greater detail later in this chapter.

Table 2.1 A layered architecture of e-commerce systems

Layer name	Services provided by layer
Application Layer	• B2C e-commerce • B2B e-commerce • C2C e-commerce
Middleman Services Layer	• Value-added networks • Public key certifying authority • Electronic payment schemes • Electronic cash • Hosting services
Messaging Layer	• Digital encryption standard • Advanced encryption standard • Public key encryption • Digital signature • Electronic data interchange • Hypertext markup language: HTML • Extensible markup language: XML
Network Services Layer	• E-mail • World Wide Web services; browsers • Hypertext transfer protocol: HTTP • Search engines
Logical Layer	• Internet • Intranet • Extranet • Firewalls
Physical Layer	• Local Area Networks • Public Switched Telephone Networks • Private Communication Networks • Optical Fibre and Coaxial Cable Networks • Satellite based Networks • Wireless Networks

We call the next layer logical layer as it defines protocols (i.e., a set of mutually agreed rules) to communicate logically between computers connected by the physical network. The Internet is a worldwide network of computers which communicate using a particular protocol known as TCP/IP (Transmission Control Protocol/Internet Protocol).

The worldwide acceptance of this standard has led to the emergence of the Internet as the essential infrastructure for e-commerce. The simplicity of connecting computers from diverse manufacturers using TCP/IP protocol led to the explosive growth of the Internet and its wide acceptance. Organizations found it attractive to use the same protocol, namely, TCP/IP to interconnect computers within their organization. A major advantage of doing it, besides allowing the organization to interconnect computers made by several manufacturers, is the availability of many services such as e-mail, file transfer protocol and browsing on the Internet to be adopted inexpensively within an organization. Such a local network within an organization is called an *intranet*. The Internet allows anyone to connect to it. It is thus vulnerable to misuse by anti-social elements who break into others' computers and steal or destroy valuable files. Special precautions are required to prevent unauthorized access. This is provided by what are known as *firewalls* which guard the intranets of organizations. Firewalls do not provide absolute security from intruders. Thus many organizations do not connect their intranet to the Internet. This, however, would prevent electronic communication among cooperating organizations. So many cooperating organizations lease communication lines and create a private network interconnecting their intranets. The protocol is, of course, TCP/IP. Such a private network interconnecting cooperating organizations is known as an *extranet*. Private network formed by leasing communication lines is expensive. Thus, methods to ensure secure communication on the Internet between cooperating organizations have been designed. This is called a Virtual Private Network (VPN). A VPN using TCP/IP protocol with enhanced security can also be called an extranet.

The next higher layer is the network services layer. This provides services on the Internet infrastructure. The Internet is similar to a railway system which is an essential infrastructure for transporting passengers. The physical layer in a railway system consists of the railway tracks, engines and carriages. The logical layer is the signalling system which specifies rules to be followed by engine drivers, guards and station masters for orderly use of the tracks by trains. If rules are broken, collisions take place. The services layer in a railway system provides reservations for passengers, facilities to transport goods, etc. Similarly, the network services layer provide services which can be carried out conveniently using the Internet infrastructure. The most important service which provides users convenient access to information stored in computers anywhere in the world is the World Wide Web. Other services which make e-commerce possible are e-mail, browsers and search engines.

Among the most important requirements of e-commerce is exchanging messages and documents between the participants in e-commerce. For example, purchase orders, delivery notes, etc., are to be sent electronically. The cheapest means of doing it is using the Internet. In B2C and C2C e-commerce the Internet is the only available system.

We require languages to compose messages which can be interpreted by computers. Hypertext Markup Language (HTML) and Extensible Markup Language (XML) provide this facility. As was pointed out earlier, the Internet being accessible to everyone there is always the danger of messages and documents being maliciously altered by unscrupulous persons. Thus, there is a need to send messages which are coded using a secret code. It is also necessary to have an equivalent of signing in the electronic medium also. Those requirements are met by the messaging layer.

We call the next layer middleman services. They are essentially services provided to e-commerce participants to make their dealings easier. In our railway analogy, a travel agent who books your railway tickets saving you a trip to the railway station provides middleman services. In e-commerce some important middleman services are secure payments using credit cards, imitating cash payments for small purchases. To authenticate digital signatures, we need an authority to certify public keys of individuals and businesses. The Government of India has created such an authority. Value-added networks provide secure electronic transactions among participants. Hosting services provide among other facilities web presence for organizations and electronic catalogues and directories to participants.

All the services provided by the layers described above are essential to support our applications, namely, B2C, B2B and C2C e-commerce. This is thus the top layer (namely, application layer) in our layered architecture.

In the rest of this book, we will describe in greater detail each of these layers and how they cooperate to provide e-commerce solutions for many day-to-day needs of persons and organizations. In the rest of this chapter, we will look at the physical layer required to transact business electronically.

2.2 LOCAL AREA NETWORK

A network connecting computers in a small geographical area such as a building, a university campus or a set of contiguous government offices is called a Local Area Network which is abbreviated as LAN. Usually computers connected to a LAN lie within a radius of about 10 km. In order to connect a computer to a network, an additional electronic circuit known as a network interface card (NIC) is connected to it. This card is required to send messages from the computer to other computers in the LAN and also to receive messages from other computers. A computer wanting to send a message will put it in a small memory (called a *buffer*) in the NIC and continue with its other tasks. The message will have a header giving the address of the destination computer, i.e., the address of the computer to which this message is to be sent. It is now NIC's responsibility to transmit the message to the specified destination. Such delegation of work by a computer allows computers with widely divergent speeds to communicate with one another in an orderly fashion. One of the earliest methods of interconnecting computers is by connecting their respective NICs to a coaxial cable as shown in Figure 2.1. This is known as *Ethernet connection*.

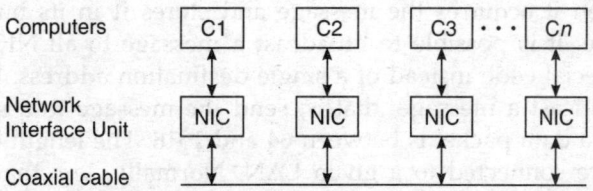

Figure 2.1 A local area network.

In order to communicate among computers connected to a common cable over which only one message can travel at a time, there is a need for a mutually agreed rule for use of the communication medium. Such a rule is called a *communication protocol*. A protocol known as CSMA/CD (Carrier Sense Multiple Access with Collision Detection) is the one used by Ethernet. Each computer connected to the Ethernet sends a message as a sequence of bits which is a train of electrical pulses. Such a transmission is known as *base band transmission*. When a computer wants to send a message it delegates the task to the NIC by placing the message in NIC's buffer. The NIC listens to the cable to find out whether any signal is being transmitted by it. This is called *Carrier Sense* (CS). If no signal is detected, it transmits a data packet. The fact that it did not sense a signal in the cable before transmission is no guarantee that there is no other message being sent in it. Messages take time to travel in the cable. A message may have been put in the cable by a distant NIC which may not have reached the interrogating NIC. The NIC may thus have wrongly assumed that there is no signal in the cable. If by chance another message was already being carried by the cable when an NIC put its message in it, the two messages will collide and both will be spoiled. Thus, the sending NIC's receiver must listen to signals in the cable for a period of time slightly larger than the time needed for a message from the most distant NIC's transmitter to reach it to detect collision. If a collision is detected by a NIC, it sends a *jamming* signal in the cable so that all NICs connected to it know that a collision has occurred. When this occurs both the NICs which had put messages in the cable must retransmit. The NIC which detected collision backs off and waits for a random time larger than T, the maximum travel time of a message between the two farthest NICs, and retransmits the message. As it waited for a random time before retransmission, the probability of another collision occurring is low. If there is a collision again, it waits for double the previous wait time and retransmits the message. By experiments and analysis, it is found that this method is quite effective, and collisionless transmission will take place soon. In fact, this method has now been used for over two decades and found extremely successful. This method of accessing the bus and transmitting packets is known as *Carrier Sense Multiple Access with Collision Detection* (CSMA/CD). It is called multiple access as any of the NICs connected to the cable can send/receive messages being sent in it—of course one at a time. Each message packet has a source and a destination address along with the message. A simplified format of a message is shown in Figure 2.2.

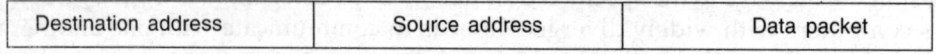

| Destination address | Source address | Data packet |

Figure 2.2 Simplified format of a message.

All NIC's receivers monitor traffic in the cable. If a NIC finds that the message is addressed to it, then it acquires the message and stores it in its buffer storage; else it ignores the message. It is possible to broadcast a message to all NICs connected to the cable by using a special code instead of a single destination address. It is also possible in this method to multicast a message, that is, send the message to a subset of NICs. The number of bytes in a data packet is between 64 and 1518. The length is decided based on how many NICs are connected to a given LAN. Normally, less than 32 computers are connected to Ethernet to reduce the probability of collisions.

We have used a coaxial cable connecting computers in a LAN in Figure 2.1. This is now obsolete and has been replaced by a connection shown in Figure 2.3. In this method of interconnecting computers, the NICs of computers are connected to what is known as a *hub* by twisted pairs of copper wire (ordinary telephone lines) (See Figure 2.3). The hub has electronic circuits which receive signals from the twisted pair connected to it, amplifies and reshapes it and broadcasts it to all the other NICs connected to it.

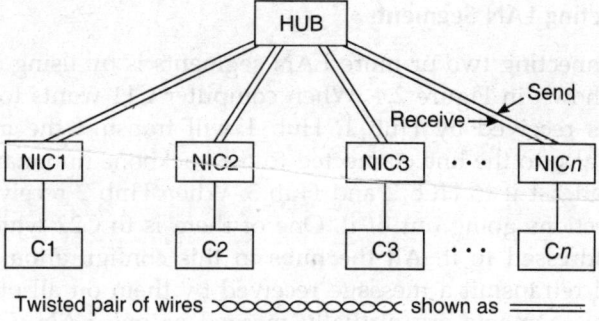

Figure 2.3 10 Base T connection of computers.

The protocol used for communication is the same as the one used before, namely, CSMA/CD. The hub detects collisions and sends this information to all the NICs connected to it. Each hub can handle from 8 to 32 computers. The distance between an NIC and the hub must be less than 100 metres. The main advantage of this type of connection compared to cable connection is higher reliability and ease of troubleshooting. Unlike a cable connection where if there is a fault in the cable, all computers connected to the LAN are affected; in a hub connection if a NIC or computer fails, it can be isolated and repaired while other computers can still work. Another advantage of hub connection is its flexibility. Adding new computers to the LAN is easy as it can be connected to the spare terminals in a hub. This connection is called 10 Base T. In this notation 10 indicates the speed of transmission of messages on the LAN in megabits per second (Mbps). Base indicates that it is base band transmission, that is, pulses corresponding to bits of a message are transmitted as they are, that is, they are not modulated. T indicates that the wire connecting the computers is a twisted pair of copper wires, similar to what is used in telephones. 10 Base T has now been upgraded to 100 Base T, that is, 100 Mbps twisted pair based LAN using a hub connection. Currently, the speed has been further increased to 1000 Mbps more commonly known as *Gigabit Ethernet*. Either twisted pair of wires or fibre optic cables are used to interconnect computers. At this high speed, the distance between the hub and the NIC with so called category 5 unshielded twisted pair of wires is 100 metres whereas with fibre optic cable it is over 250 metres.

We have seen so far what is known as a LAN segment, that is, a set of computers connected by using one hub typically in one department. The number of computers in such a LAN segment will be less than 32 and the maximum distance will be restricted to around 100 metres. Organizations are made up of a number of departments which may be in different floors of one building or may be dispersed as in a campus. Each department will have its own LAN segment. These LAN segments should be connected to enable communication between departments. For example, a purchase office may have its own LAN with 16 computers and an accounts office another LAN with 24 computers. After a purchase order is executed and goods are taken into stock, the purchase office must inform the accounts office that the vendor's bill may be paid. This information is normally sent from a computer in the purchase office LAN to one in the accounts office LAN which is possible only if the two LANs are interconnected.

2.2.1 Interconnecting LAN Segments

One way of interconnecting two or more LAN segments is by using another hub (called a *backbone hub*) as shown in Figure 2.4. When computer C11 wants to send a message to C22, the message is received by Hub 1. Hub 1 will transmit the message to all lines connected to it and also to the line connected to the backbone hub, namely, Hub BB. Hub BB in turn will broadcast it to Hub 2 and Hub 3. When Hub 2 receives the broadcast, it puts it on all connections going out of it. One of them is to C22 which will pick up the message as it is addressed to it. All the hubs in this configuration are simple as they merely amplify and retransmit a message received by them on all other connections. In this design all LAN segments are virtually merged as one LAN. Collision probability increases as the number of computers connected to Hub BB increases.

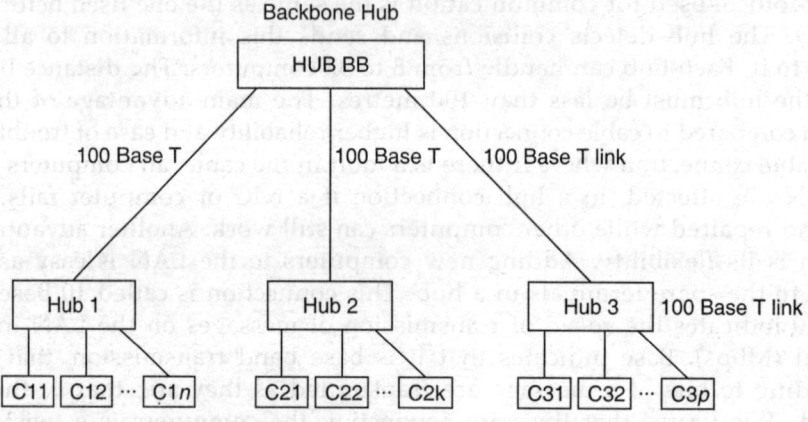

Figure 2.4 Use of backbone hub.

As long as the total number of computers connected to the backbone is smaller than about 40, this method is satisfactory.

When the total number of computers to be connected using a backbone connection is larger than 40, use of a backbone hub may lead to too many collisions and consequent reduction of speed of communication between computers. In such a case instead of using a backbone hub to connect LAN segments we use what is known as a *backbone bridge* (Figure 2.5). A bridge has features to examine each message packet which arrives and determines its destination address. It stores a table which tells the addresses of computers on each of the hubs. Using this information the bridge routes the message to the appropriate hub. A bridge thus isolates the LAN segments controlled by different hubs, thereby, eliminating collisions between messages sent in each of the LAN segments. In theory a bridge can connect any number of hubs provided it has enough memory to store address table and message being forwarded. Another advantage of a bridge is that it can control hubs working at different speeds. For example, Hub 1 may be a 10 Base T and Hub 2 100 Base T. The speeds of individual LAN segments is not relevant as a bridge effectively isolates them by storing and forwarding messages.

Infrastructure for E-Commerce **23**

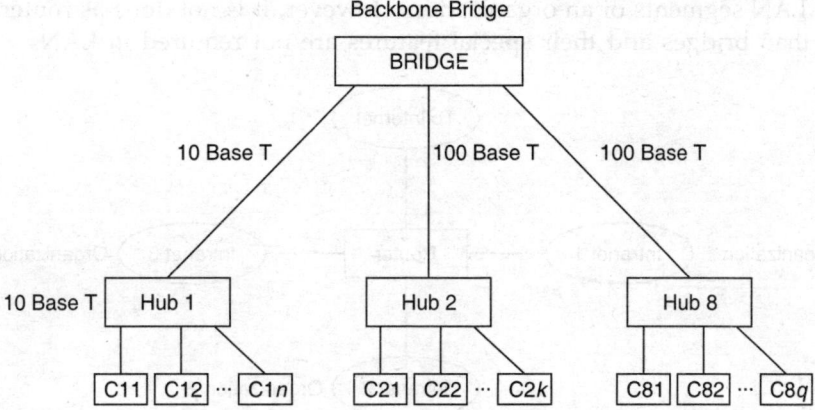

Figure 2.5 Use of backbone bridge.

A third method of interconnecting LAN segments and/or independent computers is called an *Ethernet switch* (also called *layer 2 switch*) (See Figure 2.6). An Ethernet switch can simultaneously send and receive messages from multiple sources. For example, while messages are being exchanged between Hub 1 and Hub 2, the mail server may be receiving mail from Hub 3 and sending messages on the Internet connection. This is called *full duplex communication*. An Ethernet switch can be configured with many ports with each port having a different speed. Thus, some ports may cater to 10 Base T traffic (10 Mbps) while others may cater to 100 Base T traffic (100 Mbps). The cost of a switch will depend upon the number of ports and the speed of each port. Normally, the cost of switches is higher than that of bridges and they will be used only when full duplex multiple independent communication among LAN segments/servers is essential.

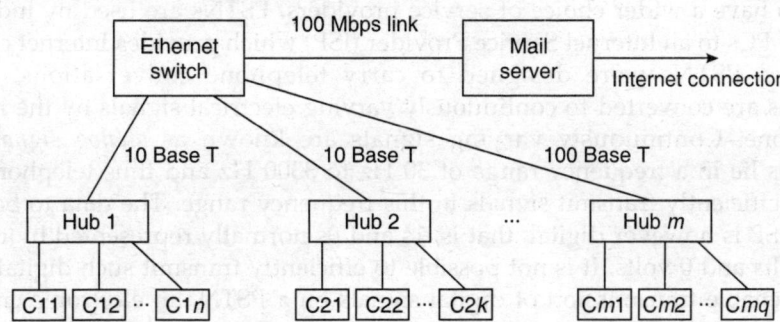

Figure 2.6 LAN segments connected with an Ethernet switch.

The last interconnection device we will discuss in this section is called a *router*. Normally routers are used to connect a company's intranet with the Internet or intranets of other companies (See Figure 2.7). In other words routers are used to connect several organizational intranets together to create a wide area network (called a WAN).

A bridge forwards messages using the physical address of a computer connected to a LAN segment whereas a router uses the IP address of a computer. As IP addresses are assigned to most computers connected to the Internet, a router can connect any two computers anywhere in the world provided they have an IP address. In theory a router can be used to connect LAN segments of an organization. However, it is not done as routers are more expensive than bridges and their special features are not required in LANs.

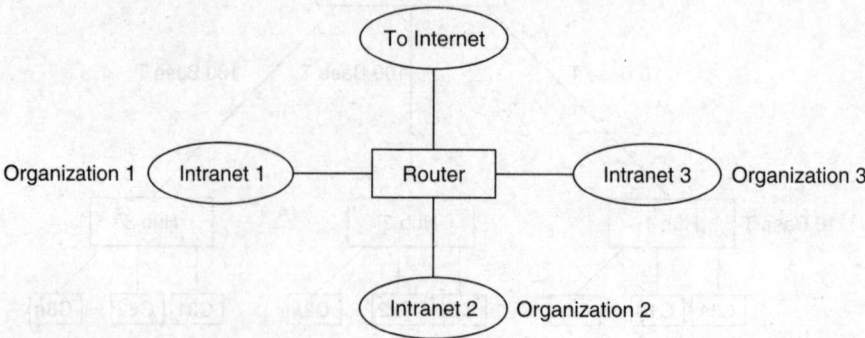

Figure 2.7 Routers are used to connect intranets and computers connected to the Internet.

2.3 PUBLIC SWITCHED TELEPHONE NETWORK

A Public Switched Telephone Network (PSTN) is the telephone networks maintained by governments or companies in many countries mainly to allow telephone communication among their customers. PSTN is also known as Plain Old Telephone Systems (POTS) in many books. In India, the Department of Telecommunication was maintaining PSTN which has now been handed over to Bharat Sanchar Nigam Ltd. (BSNL). Besides BSNL, MTNL (Mahanagar Telephone Nigam Ltd.), a government company, manages the PSTN of Delhi, Mumbai, Chennai and Kolkata. In the last few years, the Government of India has also licenced many private companies to create and manage PSTNs of their own allowing customers to have a wider choice of service providers. PSTNs are used by individuals to connect their PCs to an Internet Service Provider (ISP) which provides Internet connectivity to customers. PSTNs were designed to carry telephone conversations. Telephone conversations are converted to continuously varying electrical signals by the mouthpiece of a telephone. Continuously varying signals are known as *analog signals*. Human conversations lie in a frequency range of 30 Hz to 3300 Hz and thus telephone lines are designed to efficiently transmit signals in this frequency range. The data to be sent from a PC to the ISP is however digital, that is, 1s and 0s normally represented by two voltage levels, +3 volts and 0 volts. It is not possible to efficiently transmit such digital signals on a PSTN. To enable transmission of digital signals on a PSTN, an electronic circuit called a *modem* is employed. Modem is an abbreviation of *modulator demodulator*. A modem is connected to the PC of a customer and another to the ISP (See Figure 2.8). The modulator converts 0s and 1s to two analog signals of different frequencies, for example, a 0 may be represented by a tone at 800 Hz and a 1 by a tone at 1200 Hz. The modem at the ISP will convert these tones back to 0 and 1. The modems and the line connecting a customer's PC with the ISP is said to be duplex as information can be transmitted in both directions, that is, from ISP to customer and from customer to ISP. There are two types of modems—one is called built-in modem. This is usually a printed circuit board connected to a PC's motherboard. The other is an external modem which is a box connected to the communication port of a PC. Early modems used to carry digital information at the rate of 1200 bits per second (bps). With advances in digital signal processing modern modems are capable of speeds up to 56 Kbps using the same old telephone lines. The telephone line connecting a home PC to the telephone exchange is called a *local loop*. It is also called a dial-up telephone line. Even though the modem may be a 56-Kbps modem, you may not get this speed in practice as the quality of telephone lines in most cities is not very good.

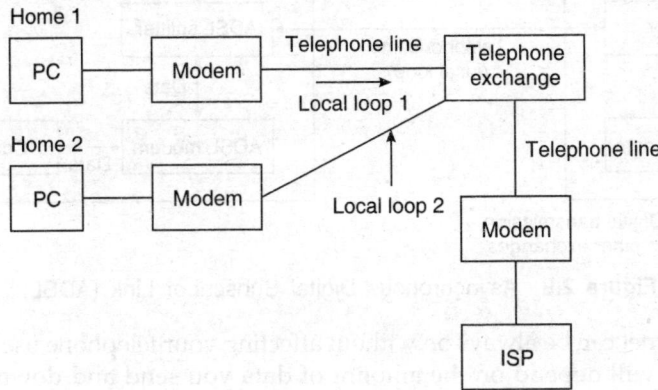

Figure 2.8 Modem connection from homes to ISP.

2.3.1 Broadband Connection to Home PC

The speed obtainable with dial-up modem connection is too low to download audio and video files using the Internet. Thus nowadays a so-called broadband connection to homes is provided by PSTNs. This technology known as ADSL (Asymmetric Digital Subscriber Link) uses a modem called ADSL modem to connect telephone line at user premises with the local telephone exchange. This technology uses the fact that copper wires used as telephone lines can inherently support a large bandwidth, much larger than that needed for voice communication. This bandwidth was reduced intentionally in voice communication to prevent noise from higher frequencies corrupting voice communication. In ADSL technology a filter called a *splitter* is used in both subscriber's premises and at the exchange. This splitter divides the bandwidth into two disjoint parts. The lower frequencies of 30 Hz to 3300 Hz is used for telephone conversation and the higher band 1 MHz to 8 MHz is used for data communication. The arrangement is shown in Figure 2.9.

At the customers premises the ADSL modem converts 0s and 1s from PC to analog signals in 1 MHz to 8 MHz band. The splitter isolates this from telephone conversations at the lower frequency band of 30 Hz to 3300 Hz. In other words, in the telephone line voice signals and digital communications from PC can be simultaneously sent as they are in two different bands which are widely separated and thus will not interfere with one another. Normally most users download lot more data than they send to ISP. Thus, data speed from home PC to ISP is usually limited to 256 Kbps and from ISP to home PC is around 1 Mbps. The most significant advantages of using ADSL modem connections are as follows:

1. Telephone conversations can go on while one is using the Internet.
2. The speeds are much higher than dial-up modem connection.

26 *Essentials of E-Commerce Technology*

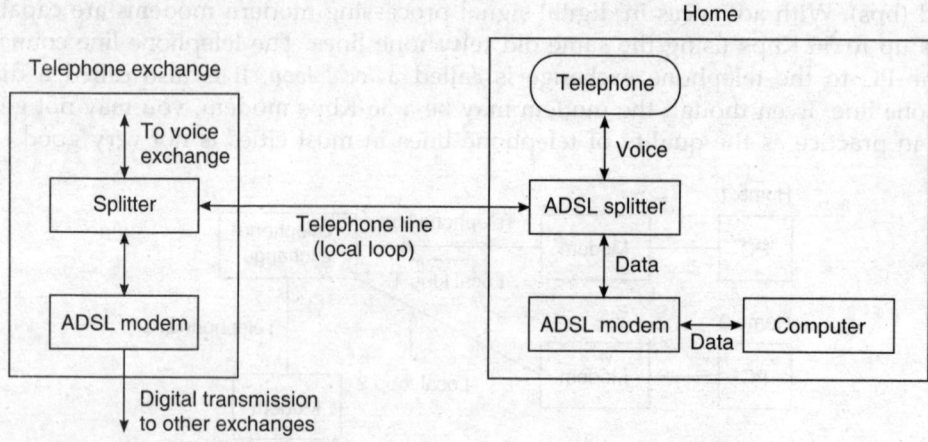

Figure 2.9 Asynchronous Digital Subscriber Link (ADSL).

3. The Internet can be always on without affecting your telephone use for conversation.
4. Charges will depend on the amount of data you send and download and not on the time the Internet is used (There is usually a fixed monthly rental charged by PSTN).

The actual speeds customers get depend on the quality of telephone lines and distance from customer premises to telephone exchange. The numbers quoted are average values.

2.3.2 ISDN Service

Another facility provided by PSTN nowadays is called an ISDN service. The expansion of ISDN is Integrated Services Digital Network. ISDNs provide all digital links between customer's premises and the telephone exchange without a modem. This network is different from the analog phone network. ISDNs are also dial-up lines, that is, a customer dials up an ISP and pays only for the time he or she is connected. ISDNs are comparatively expensive and normally only organizations with large traffic use them. ISDNs normally provide data rates ranging from 128 Kbps to 2 Mbps. Dial-up modems and ISDN networks are widely deployed. In India, ISDN is now available from BSNL and other providers in many larger cities. Due to higher bandwidth available with ISDN, they are effective for video transmission such as videoconferencing.

2.4 CABLE NETWORK

Another technology which does not use the telephone network but the cable TV network for Internet traffic is emerging. The penetration of cable TV in India is double the penetration of telephones. In other words, the number of people who have cable TV connection is double those who have telephone connection. Thus, it is a good alternative communication system for e-commerce. In the cable TV system, a cable operator broadcasts satellite TV programmes using coaxial cables and amplifiers to homes. As illustrated in Figure 2.10 fibre optic cables connect a major operator, who receives all satellite channels, to franchisees who are in various localities of a city. The distributor runs coaxial cables between his premises (called *cable head end*) and hundreds of homes in the locality. The cable network is a tree network with several branch networks. (In India they are literally tree top networks!). The cable network is in fact a one-way communication network transmitting analog video signals to customers. If it is to be used in e-commerce, there must be two-way digital communication. This is achieved by a device called a *cable modem*. Each home user should buy a cable modem and connect it to his/her PC to use this service. Cable modem is a box which is connected to a PC via 10 Base T Ethernet port. This modem divides the cable network into two channels, a downstream channel from the ISP (connected to the cable operator) to homes and an upstream channel from homes to the ISP. The downstream channel has a high bandwidth, around 10 Mbps and the upstream channel a bandwidth of around 768 Kbps. Unlike ADSL where one telephone line connects the exchange to each home the cable network is a shared network as is clear from Figure 2.10. Thus, the bandwidth is shared by several users and the effective bandwidth, if everyone simultaneously downwards video or surfs the web, will be small. The upstream traffic also shares the same cable network and packets sent by different users on the same network will collide. Protocols similar to Ethernet protocol is thus required in the cable systems. The cable medium cost is also much higher than a telephone modem cost (about 5 times). As of now use of PSTN with ADSL is more popular.

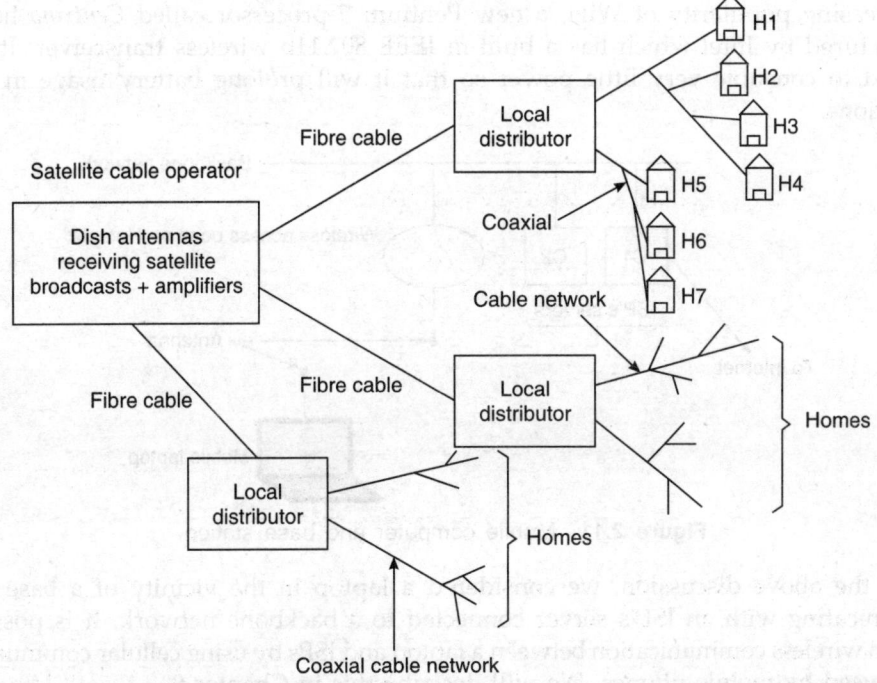

Figure 2.10 Cable TV network using fibre connections to local distributors and coaxial cables from distributor to homes. Each distributor handles several hundred homes.

2.5 WIRELESS NETWORKS

Wireless networks are becoming important in e-commerce as customers often want to order goods while they are travelling, that is, while mobile. Goods in movement are often tracked to determine their time of delivery or sometimes to modify their delivery location. These are called *mobile commerce applications* or *M-commerce* and depend on wireless networks. In India the number of users of mobile phones is much larger than the users of PC. So e-commerce using mobile phones is becoming very popular. In this section, we will restrict our discussion to wireless network which use PCs. Mobile phone networks will be described later in Chapter 8 of this book.

Wireless technology is primarily used to communicate between mobile laptop computers and ISPs connected to a backbone high-speed network (usually fibre optic cable). In order to use wireless communication, a mobile laptop computer should have a built-in wireless transceiver (a combination of a transmitter and a receiver) and the backbone must have a wireless access point with a transceiver to transmit and receive data from the mobile computer (See Figure 2.11). The transmitter uses a frequency in the so called unlicensed 2.4 GHz frequency band which is not used for commercial radio and other purposes. The currently popular wireless connection is called WiFi (Wireless High Fidelity). It uses a standard called IEEE802.11b (standardized by IEEE, U.S.A.). It uses 2.4 GHz frequency band and can transmit and receive data at the rate of 11 Mbps. The distance from mobile laptop to base station should be less than 30 metres. Base station called "hotspots" are appearing in hotels, airports, large stores and even street corners in some big cities. With the increasing popularity of WiFi, a new Pentium 5 processor called *Centrino* has been manufactured by Intel which has a built-in IEEE 802.11b wireless transceiver. It is also designed to consume very little power so that it will prolong battery usage in mobile applications.

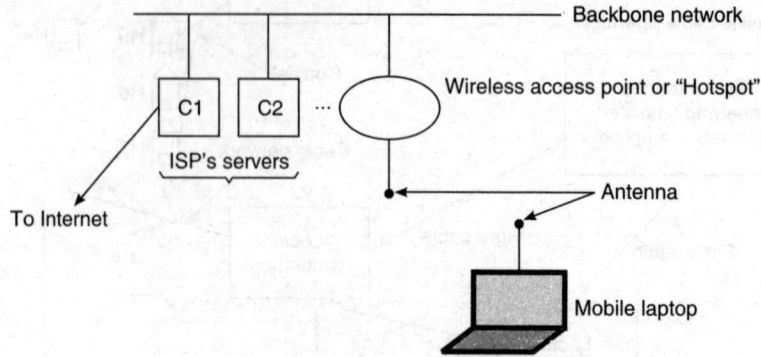

Figure 2.11 Mobile computer and base station.

In the above discussion, we considered a laptop in the vicinity of a base station communicating with an ISP's server connected to a backbone network. It is possible to maintain wireless communication between a laptop and ISPs by using cellular communication system used by mobile phones. We will describe this in Chapter 8.

2.6 MICROWAVE AND SATELLITE NETWORK

For high bandwidth communication over long distances microwave and satellite communication systems are used. Typically, microwave links are used to communicate between an organization and an Internet Service Provider for wide band communication.

For example, a company called Software Technology Park of India (STPI) provides satellite links to overseas customers to a number of software export companies in towns such as Bangalore, Chennai, Hyderabad, and Mumbai. In this case, each software exporting company sets up a microwave link from their building (normally on their rooftop) to STPI. STPI provides a satellite communication link from its premises to overseas communication companies which receive data transmitted and distribute them to companies located in their country (or neighbouring country).

Microwave links use the frequency band from 2 to 40 GHz. At these frequencies called microwave frequencies the electromagnetic waves cannot bend or pass through obstacles such as buildings or hills. The transmitter and receiver should be in a line of sight without any obstacles in between. That is the reason the microwave transmitters and receivers, which are small dish-like structures (See Figure 2.12), are mounted on top of tall buildings in cities. Inter-city microwave links are set up over hilltops (which you often see from trains). Microwaves are attenuated (i.e., loose their strength) during transmission. For transmission within a city where the distance between a transmitter and a receiver is less than 50 km this is not a problem. If the distance is more we need what are known as microwave repeater stations. A repeater is placed at a distance of around 50 km from a transmitter. This repeater receives the microwave signal amplifies it and retransmits it. (See Figure 2.12). The major advantage of microwave transmission is the large available bandwidth for signals permitting data transmission rates up to 250 Mbps. The capital investment required is high—around Rs. 5 lakhs for a link within a city and around Rs. 20 lakhs for intercity repeater station. Thus, they are used only when the data traffic to be transmitted/received is very high. For instance, a microwave link can support 250,000 digital channels each capable of transmitting/receiving 1 Kbps.

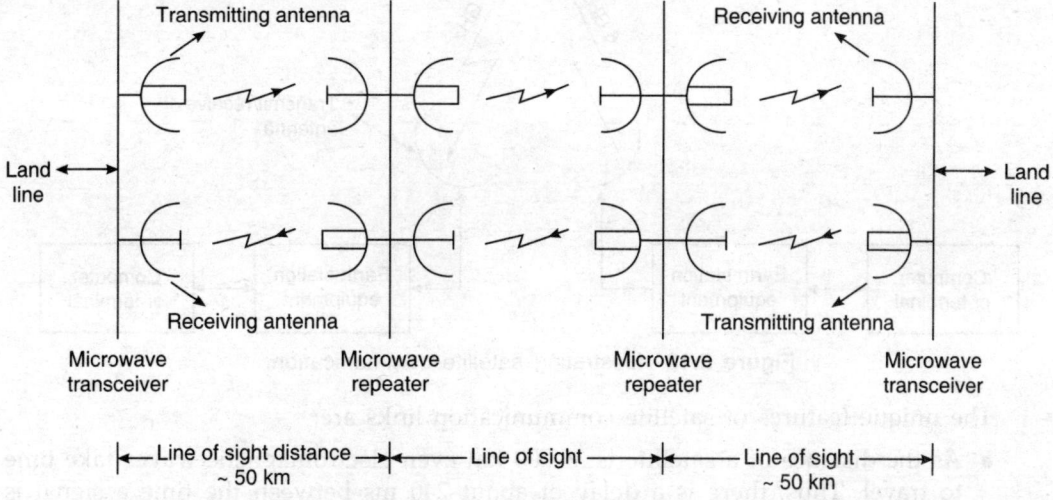

Figure 2.12 Microwave relay.

2.6.1 Satellite Communication

Communication satellites are very useful to send and receive high bandwidth data communications between widely separated points. For example, satellite links operate between India and Europe. Communication satellites are now launched either by launch vehicles (rockets) or by space shuttles. Satellites are parked in a *geostationary* orbit at 36,000 km above the equator. The speed of a satellite in this orbit equals the speed of rotation of the earth and thus it is stationary relative to earth. The Indian National Satellite known as INSAT 2D is parked in an orbit so that it is "visible" from any place in India including Andaman, Nicobar and Lakshadweep.

A communication satellite is essentially a microwave relay station in the sky. Microwave signal at a frequency of 6 GHz carrying the data is transmitted to it from an earth station which has a transmitter. The signal travels a distance of 36,000 km to the satellite and is received as a feeble signal by a system called a transponder mounted on the satellite. It is amplified and retransmitted to the earth using a frequency of 4 GHz by the transponder. The retransmission frequency is different as otherwise the strong return signal will mask the signal received from the earth. The main advantage of a satellite is that it is visible from anywhere in its coverage area which in the case of INSAT is India. Thus, the transmitted signal can be received at any place in India. The bandwidth which can be handled by a transponder is about 36 MHz which would support 400 digital channels each of speed 64 Kbps.

A satellite has several transponders thus providing enormous data transmission capability at a cost which is competitive to microwave links. Figure 2.13 illustrates a satellite communication link.

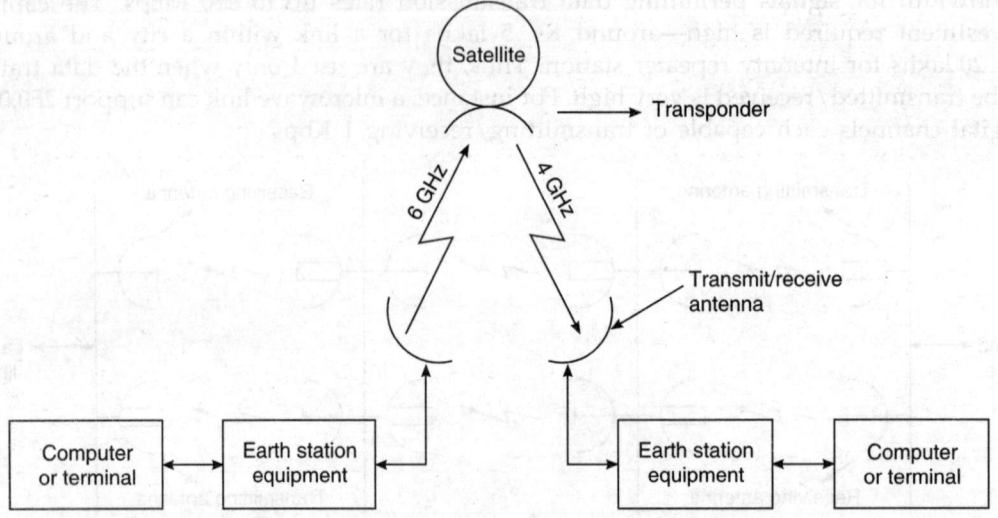

Figure 2.13 Illustrating satellite communication.

The unique features of satellite communication links are:
- As the distance of a satellite is 36,000 km even electromagnetic waves take time to travel. Thus, there is a delay of about 240 ms between the time a signal is transmitted to the time it is received by a receiver. This delay has to be accounted for in designing systems based on satellites.
- Users can install a receiving dish antenna to receive signals broadcast from a satellite at their premises. Satellite digital radio broadcasts and television broadcasts are now available and they can be received by very small dishes (around 20 cm diameter for radio and 1 m diameter for TV reception) as their bandwidth is small.

Satellite radio stations provide data links of 64 Kbps for broadcasting digital data from an organization to multiple recipients.
- A transmitting station can receive back the signal sent by it and verify whether it has been correctly transmitted and received. If an error is detected, the signal is retransmitted.
- Recently transponders in a satellite allow anyone to connect to it. It has thus been improved allowing lower cost transmitting and receiving systems and antennas to be located on the rooftops of cooperating organizations allowing them to operate private networks. These are called VSATs (Very Small Aperture Terminals). The term *aperture* refers to the diameter of the dish antenna which is around a metre. Many banks in India operate private VSAT networks to communicate between regional offices and head office. VSATs are particularly useful for connecting remote locations such as cities on hills, islands, etc. VSAT networks have also been used to transmit live lectures from an institution to several remotely located lecture halls across India.
- The frequency range of 4 GHz to 6 GHz is called the C band. A higher frequency band of 12 GHz to 14 GHz called the K_u band is the one used by many VSAT networks in India.

2.7 PRIVATE COMMUNICATION NETWORKS

We saw that PSTNs are maintained by BSNL in India and are primarily intended to cater to telephone conversations. We also saw that they are upgrading them to provide data services of which Integrated Services Data Network is the most important one. PSTNs by their very nature lend themselves to intrusion by hackers. (After all, it is meant to be used by anyone who is a subscriber). Thus, many organizations which require secure communication prefer to set up their own communication systems. For example, many banks have set up their own communication system connecting their branches. Railways, for example, have their own private network for their reservation system. The railway's private network connects almost all major stations which may be of the order of 300 spread all over India. Many major companies with branches spread all over India have also set up their own private networks. Private networks may be set up by leasing communication lines (may be voice or data lines) from a communication company such as BSNL. Private companies are also allowed now in India to set up PSTNs and they will also be in a position to lease communication lines to customers.

Private networks are based on leased lines or VSAT network. VSAT networks have the advantage that they can reach remote areas including hilly areas where PSTNs, do not have large bandwidth access networks. Private networks are expensive to set up. Thus, only large organizations can afford them. Sometimes several organizations cooperate and share the cost particularly if they all will benefit from it. For example, several smaller banks may cooperate to set up a private bank network. An automobile company may cooperate with all their major suppliers and set up a private network.

In this chapter, we have discussed primarily the lowest layer of relevance to e-commerce, namely, the physical network. It is important to understand this layer to appreciate why certain design decisions are taken in the higher layers of relevance to e-commerce. In the next chapter, we will discuss in some detail the logical layers, namely, the Internet, intranet, extranet and the need for firewalls to protect intranets from intruders.

SUMMARY

1. It is a good idea to examine any complex system as consisting of layers—each layer providing a service.
2. In a layered architecture, higher layers use the functions provided by lower layers.
3. One possible layered architecture for e-commerce consists of six layers which are:
    (i) Physical—Lowest Layer
    (ii) Logical
    (iii) Network services
    (iv) Messaging
    (v) Middleman services
    (vi) Applications—Highest Layer
   The components of each layer are given in Table 2.1.
4. The physical layer consists of the basic physical infrastructure such as local area networks, wide area networks, and wireless networks which are required for e-commerce.
5. The logical layer defines the protocols to communicate logically between computers connected by a physical network.
6. The Internet is a worldwide computer network which interconnects a large number of smaller local computer networks. A protocol called TCP/IP (Transmission Control Protocol/Internet Protocol) is used to communicate between computers connected to the Internet.
7. An organization-wide network of computers communicating using TCP/IP protocol is called an intranet.
8. A private computer network connecting the intranets of cooperating organizations is called an extranet.
9. The network services layer provides services such as e-mail, markup languages, browsers etc., necessary for e-commerce.
10. The messaging layer consists of services to enable exchange of messages between business partners in e-commerce securely. It has features to encrypt messages and to digitally sign documents.
11. The middleman services layer provides services such as electronic payment schemes, public key certification, etc.
12. The topmost application layer deals with various modes of e-commerce such as Business to Customer, Business to Business, Customer to Customer e-commerce.
13. Local Area Network (LAN) is part of the physical layer. A network connecting computers in a small geographical area, normally, within an organization is called a LAN.
14. Ethernet is a very popular LAN. Early implementation of Ethernet used a multidrop coaxial cable to interconnect computers and was designed to communicate at 10 to 100 Mbps. It uses a protocol called CSMA/CD (Carrier Sense Multiple Access with Collision Detection) to communicate between computers connected to the LAN.

15. Recent Ethernet LAN implementations use unshielded twisted pairs of copper wires to interconnect computers to what is known as a Hub. Such an Ethernet connection allows easier implementation, upgradation and trouble shooting.
16. A device known as a bridge is used to connect the hubs of a number of LANs.
17. Another device called an Ethernet switch (also called layer 2 switch) may also be used to connect the hubs of a number of LANs. A switch can simultaneously send and receive messages from a number of hubs.
18. A router is used to connect an organization's network with that of other organizations situated at geographically dispersed locations.
19. A Public Switched Telephone Network is a telephone network normally maintained by governments in many countries. In India, a government owned company BSNL (Bharat Sanchar Nigam Ltd.) has the most extensive telephone network. Other private companies are now allowed to own and operate Public Switched Telephone Networks for its customers.
20. In order to use a PSTN to send digital signals, a device called a modem is needed. A modem converts digital signals to continuously varying analog signals for transmission using PSTNs and reconverts them to digital form at the receiving end for use by computers. Modern dial-up modems provide communication speeds up to 56 Kbps using PSTN. Another system called Asymmetric Subscriber Digital Link (ADSL) uses a modem and a splitter to allow broadband digital signals (>256 Kbps) to be sent on telephone lines.
21. PSTNs have established fully digital networks called ISDN (Integrated Services Digital Network). They provide high data rates up to 128 Kbps.
22. Cable TV networks provide cables carrying video programmes to homes. Using cable modems and what is known as a set-top box, these cables can be used to send and receive digital signals. They use high speed (up to 10 Mbps) for receiving data.
23. Communication between a mobile computer such as a laptop computer and a backbone network connected to an ISP is established using wireless communication. Wireless transceivers are installed on laptop machines and also at wireless access points connected to the backbone network. The mobile computer communicates with the wireless access point which in turn communicates with ISP connected to the backbone network.
24. Wireless cellular networks are used for communication between participants in what is known as mobile commerce (m-commerce). We will discuss this in Chapter 8.
25. Microwave links are used to communicate high speed digital data over a distance of a few kilometres (line of sight). For longer distance transmission, repeater stations are used.
26. Satellite communication is used to communicate between computers located at widely separated geographical areas, namely, between cities. Communication satellites are microwave relay stations in the sky. Transponders on the satellite are used to receive, amplify and retransmit signals received from one earth station to another earth station.

34 *Essentials of E-Commerce Technology*

27. Large organizations such as the railways, banks, etc., maintain their own wide area network to facilitate communication between widely dispersed locations. Such networks usually lease private lines from PSTNs or set up their own satellite based networks using VSATs (Very Small Aperture Terminals). Such networks provide better security as they are not connected to PSTNs.

EXERCISES

2.1 What do you understand by the term "layered architecture"? Why is this idea used in describing e-commerce?

2.2 What are the different layers in e-commerce architecture? Briefly describe the functions of each of the layers.

2.3 What is the physical layer in e-commerce architecture? What are the different parts of the physical layer?

2.4 What is the communication protocol used by the Internet? What is the advantage of using such a common standard protocol to interconnect computers?

2.5 What is the Internet? What is the importance of the Internet in e-commerce?

2.6 What is intranet?

2.7 What is extranet?

2.8 What is a Virtual Private Network? In what way is it superior to extranet?

2.9 What is a LAN?

2.10 What is NIC and why is it needed?

2.11. What is Ethernet? What is the communication protocol used by Ethernet?

2.12 Explain how CSMA/CD works.

2.13 What is a hub? How is it used to configure an Ethernet LAN?

2.14 What do you understand by the terminology 10 Base T LAN?

2.15 What is a bridge? How is it different from a hub? When is it used?

2.16 What is an Ethernet switch? How is it different from a bridge? What are its advantages (if any)?

2.17 What is a router? How is it different from an Ethernet switch?

2.18 What is PSTN? Are PSTNs analog or digital?

2.19 What is a modem? When is it used? How does it work?

2.20 What is the speed of digital transmission on PSTN using modems?

2.21 What is a local loop? Is it analog or digital?

2.22 What is ADSL? How does it function? What is its use? What is the speed of digital transmission with ADSL?

2.23 How can TV entertainment cable network be used to connect home PCs to ISP?

2.24 What is WiFi? What frequency band does it use? What is the maximum speed of digital data transmission using WiFi?

2.25 When is microwave transmission used? What is its frequency range?

2.26 When is satellite transmission of data useful? Describe how a communication satellite receives and transmits data from earth stations.

2.27 What is a transponder? Normally how many 64 Kbps channels are supported by a transponder?

2.28 What is a VSAT? What are its applications?

2.29 Which types of organizations prefer to maintain their own private network? What are the advantages of a private network?

OBJECTIVE QUESTIONS

Each question has four possible answers. Pick the most appropriate answer.

2.1 A layered architecture is used to describe e-commerce because it
   (a) Is a convenient notation
   (b) Is a universally accepted standard
   (c) Provides a logical framework to describe e-commerce technology
   (d) Gives guidelines to develop e-commerce sites

2.2 In this book the number of layers used to describe e-commerce architecture is
   (a) 7   (b) 6
   (c) 5   (d) 4

2.3 In the layered architecture
   (a) Each layer requires the services offered by the layers below it
   (b) Each layer is independent
   (c) Each layer requires the services offered by the layers above it
   (d) All layers are mutually interdependent

2.4 In the layered architecture for e-commerce, the physical layer deals with
   (a) Communication systems/devices used to interconnect computers
   (b) Computer systems and their architecture
   (c) Internet for e-commerce
   (d) World Wide Web

2.5 In the layered architecture for e-commerce, the logical layer deals with the
   (a) Logic of e-commerce
   (b) Logical thinking required to design e-commerce sites
   (c) Logic of interconnecting computers
   (d) Protocols needed to communicate between computers connected to the physical network

2.6 The network services layer in e-commerce architecture deals with
   (a) Various services which can be carried out conveniently using the Internet
   (b) Services provided by e-shops
   (c) Protocols used to connect service networks
   (d) The description of the World Wide Web

2.7 The messaging layer in e-commerce architecture deals with
   (a) Messages to be sent between participating businesses
   (b) Exchanging documents and message securely between participating entities
   (c) Digital encryption standards in e-commerce
   (d) Authenticating messages transmitted between businesses

2.8 The most popular methods of interconnecting computers in a LAN is
 (a) A token ring network (b) An Ethernet
 (c) An extranet (d) A store and forward network

2.9 To connect a computer to the Ethernet, you need a
 (a) Bridge (b) Router
 (c) Network Interface Card (d) Hub

2.10 The protocol used by Ethernet LAN is
 (a) CSMA/CD (b) CSMA/CA
 (c) CDMA (d) CSAM/DC

2.11 A LAN segment usually connects
 (a) 2 to 4 computers (b) 8 to 16 computers
 (c) 48 to 64 computers (d) 128 to 256 computers

2.12 To form an Ethernet, computers are interconnected using a
 (a) Bridge (b) Hub
 (c) Router (d) Backbone hub

2.13 Two LAN segments may be connected using a
 (a) Modem (b) Hub
 (c) Router (d) Backbone hub

2.14 Several LAN segments may be connected using
 (i) A router (ii) A hub
 (iii) An Ethernet switch (iv) A bridge
 (v) An Ethernet
   (a) (iii) and (v) (b) (i)
   (c) (ii) (d) (iii) and (iv)

2.15 LANs of organizations located in widely separated areas are connected using
 (a) A bridge (b) A router
 (c) An Ethernet switch (d) PSTN

2.16 The expansion of PSTN is
 (a) Public Switched Telephone Network
 (b) Private Services Telecom Network
 (c) Privately Serviced Telecom Network
 (d) Public Sector Telephone Network

2.17 To connect a home computer to ISP using PSTN, we need a
 (a) Router (b) Modem
 (c) Network Interface Card (d) Telephone jack

2.18 By ISP we mean
 (a) Internet Service Point (b) Immediate Solution Provider
 (c) Internet Solution Provider (d) Internet Service Provider

2.19 Currently available dial-up modems carry digital data on PSTN at the rate of
 (a) 1200 bps (b) 1 Mbps
 (c) 56 Kbps (d) 512 Kbps

2.20 Data transmission using modems is
 (a) Half duplex (b) Simplex
 (c) Asymmetric (d) Full duplex

2.21 With ADSL modems, it is possible to download data from ISP at the maximum rate of
 (a) 256 Kbps (b) 10 Mbps
 (c) 1 Gbps (d) 4 Kbps

2.22 ADSL uses as transmission medium
 (a) TV cable
 (b) Telephone line and a separate data line
 (c) Telephone line connected to a subscriber's premises
 (d) Wireless communication

2.23 An ADSL system consists of
 (a) A splitter and an ADSL modem connected to a telephone line
 (b) A pair of splitter-modem set, one at the subscriber's premises and the other at the exchange
 (c) A pair of ADSL modems
 (d) A pair of splitters

2.24 A splitter used with ADSL
 (a) Splits telephone line into two lines
 (b) Is a filter to filter out noise in telephone line
 (c) Is a filter which separates audio signals used in conversation from the data signals
 (d) Allows both conversation and data to be transmitted on the same telephone line

2.25 Wireless networks are used in
 (a) e-commerce (b) m-commerce
 (c) em-commerce (d) mn-commerce

2.26 The distance of a wireless enabled laptop computer from a WiFi hotspot should be
 (a) More than 100 metres (b) Less than 1 kilometre
 (c) Less than 1 metre (d) Less than 100 metres

2.27 A microwave link is used when
 (a) High bandwidth (250 Mbps) communication over long distances is required
 (b) Low bandwidth (100 Kbps) communication over long distance is required
 (c) High bandwidth (250 Mbps) communication over short distance (<1 km) is required
 (d) Low bandwidth communication (< 100 Kbps) over short distance (<1 km) is required

2.28 In long distance microwave communication, repeaters are line of sight and placed at a distance of every
 (a) 10 km (b) 50 km
 (c) 150 km (d) 1 km

2.29 The orbit of a geosynchronous satellite used for communication is at a distance of
 (a) 36000 km from the earth
 (b) 360 km from the earth
 (c) 36 km from the earth
 (d) 72000 km from the earth

2.30 In a satellite communication, the transponder receives signals from earth at frequency x and transmits back an amplified signal at frequency of y. The values of x and y are:
(a) $x = 6$ GHz, $y = 4$ GHz
(b) $x = 4$ GHz, $y = 4$ GHz
(c) $x = 6$ GHz, $y = 6$ GHz
(d) $x = 16$ GHz, $y = 20$ GHz

2.31 Expansion of VSAT is
(a) Virtual Satellite
(b) Virtual Satellite Advanced Technology
(c) Very Simple Architecture Terminal
(d) Very Small Aperture Terminal

2.32 Private computer communication network can be set up using
 (i) VSATs
 (ii) Leased telephone lines
 (iii) TV cable network
 (iv) Internet
(a) (i) and (iii)   (b) (i) and (ii)
(c) (ii) and (iii)   (d) (i), (ii), (iii), (iv)

CHAPTER 3
Communication Networks for E-Commerce

LEARNING GOALS

In this chapter, we will learn:
1. How the Internet works.
2. About the protocol used by the Internet and its importance.
3. About firewalls needed to protect e-commerce sites and customers from unauthorized access.

3.1 INTRODUCTION

In the last chapter, we examined the physical layer, namely, the communication systems used to interconnect computers. In particular, we saw how computers are connected as a Local Area Network and how individual computers can send digital information using a Public Switched Telephone Network. E-Commerce depends on individuals and businesses to be able to communicate using a worldwide communication infrastructure. Such a worldwide communication infrastructure is provided by the Internet. In this chapter, we will explain how the Internet functions.

3.2 THE INTERNET

We saw that organizations connect computers within their premises as a Local Area Network. The main purpose of interconnecting computers is for resource sharing. When computers can easily communicate with one another, they can share resources such as files. Expensive peripherals connected to one computer in the network can be used by any other computer. A network may be organized in such a way that some specialized services offered by one machine may be used by other machines. The machine offering special services is called a *server* and those using this service are called *clients*. For example, there may be a machine in an organization called a *database server* which may store all the important data needed by an organization. Another machine may be a high-speed processor and serve as a *compute server*. Yet another machine may provide mailing services which allows one to send and receive messages. In other words, computers connected to a LAN may be thought of as several servers each providing a specified service and several clients using these services. The servers as well as the clients may be from several different manufacturers. To enable computers to cooperate, it is necessary to have a software system and a commonly agreed upon set of rules for communication which would allow them to exchange data and commands. Such a set of rules is called a *communication protocol*. Thus, a communication protocol enables computers manufactured by different vendors having diverse hardware architecture to function harmoniously. This idea was extended to enable connection of several LANs which are geographically dispersed all over the world to form a giant network which can exchange messages and share resources. Such a connection of computers distributed all over the world constitute what is known as the *Internet*. Physical connection is necessary to create a network of computers but a common communication protocol is vital. Such a common protocol has evolved and is used by all computers connected to the Internet. This is called *Transmission Control Protocol/Internet Protocol* (TCP/IP for short). Thus, the Internet has to be described at two levels. The lower level is the method of physically interconnecting LANs spread out all over the world and transmission of data from one machine in one LAN to another machine in another LAN situated in another corner of the world. At a higher level is the commonly understood and agreed upon rules of communication, namely, the communication protocol. It is important to understand that the communication protocol is implemented as a software in each computer. The software which runs on the computer is designed so that the communication of data is independent of the hardware architecture. We will first examine the lower level, namely, the physical interconnection.

3.3 PACKET SWITCHING AND THE INTERNET PROTOCOL

In Figure 3.1, we show a physical interconnection of computers constituting the Internet. Observe that in this figure, we have shown five LANs. For two LANs in two organizations to communicate with one another, there is a need for a hardware device called a *router* which connects them. The router is actually a special purpose computer (with appropriate software) whose task is to receive data from one LAN and forward it to the other LAN connected to it. For example, if a computer C1 connected to LAN1 wants to send data to C4 in LAN2, LAN1 first finds out whether C4 is connected to LAN2. If it is then the data from C1 is sent to the router R1 which connects LAN1 and LAN2. Router R1 forwards the data to C4 in LAN2. Routers can receive data from either direction and forward them as requested. In other words, R1 can be used to send data from LAN1 to LAN2 or from LAN2 to LAN1.

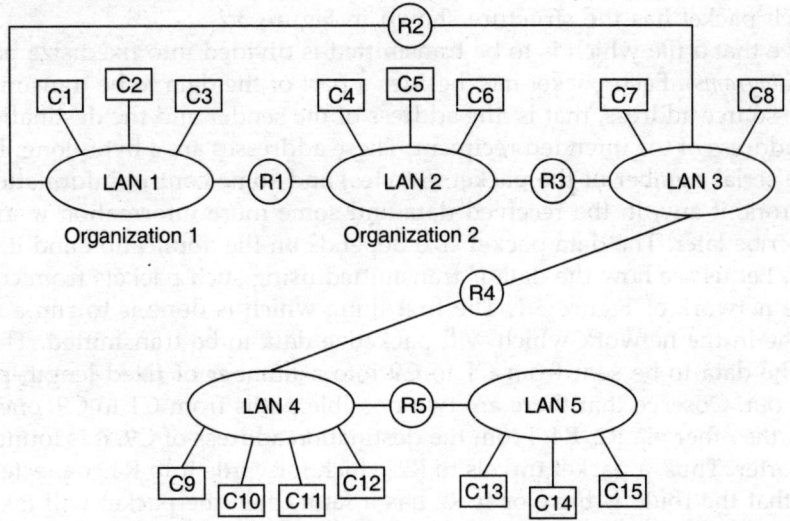

Figure 3.1 Connecting LANs of several organizations using routers.

In order for a router to forward data from a computer connected to a LAN to another computer in another LAN it needs to know the identity of the destination computer. Identity of a computer connected to the Internet is uniquely determined by what is known as its *address*. The address has been standardized as a string of 4 bytes (32 bits). Thus, a total of 2^{32} = 4 billion addresses are available and, in theory, it is possible to connect 4 billion computers to the Internet.

The second important point to remember is that while C1 sends data to C4, another machine C2 on LAN 1 may like to send data to C6. As there is only one communication path between LAN1 and LAN2 via R1, it implies that C2 has to wait for C1 to complete its data transmission. If the data is a large file, say F1, the line will be busy for a long time. Even if the data to be sent by C1 is a short item still it has to wait. This is not fair. Further as F1 is long the probability of an error occurring in transmission is high, and F1 may have

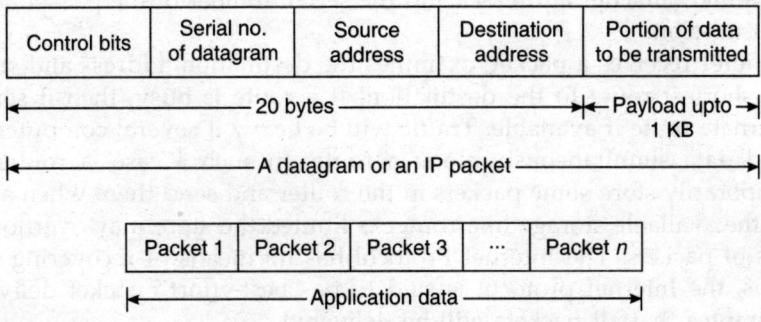

Figure 3.2 Application data broken into packets.

to be retransmitted if an error is detected. In order to alleviate both these problems application data to be transmitted from one computer to another in the Internet is broken up by the software running on the machine into a number of smaller chunks called *packets* and transmitted. In other words, a file to be transmitted is divided into a number of fixed size packets. Each packet has the structure shown in Figure 3.2.

Observe that a file which is to be transmitted is divided into fixed size packets (also known as *datagrams*). Each packet has, besides a part of the data to be transmitted (called *payload*), the source address, that is, the address of the sender and the destination address, that is, the address of the intended recipient. These addresses are 4 bytes long. Each packet also has the serial number of the packet (8 bytes) and some control information (4 bytes) to detect errors, if any, in the received data and some more information whose purpose we will describe later. The data packet size depends on the application and its maximum size is 1 KB. Let us see how the data is transmitted using such packets from computer C1 to C9 in the network of Figure 3.1. The first thing which is done is to run a software in each machine in the network which will packetize data to be transmitted. This software breaks up the data to be sent from C1 to C9 into a number of fixed length packets and sends them out. Observe that there are two possible paths from C1 to C9, one via R1, R3 and R4, and the other via R2, R4. From the destination address of C9, it is found that route via R2 is shorter. Thus, a packet travels to R2 which forwards it to R4 connected to LAN4. If suppose that the route is busy or if R2 has a fault, then the packet will take the route R1, R3, R4 and reach C9. Individual packets belonging to a long message may arrive out of order at the destination. The serial numbers of the packets into which a message has been broken up are necessary to reassemble a message at its destination. The communication protocol described above which is implemented as software on every computer is called the *Internet Protocol*. The protocol has the following important features:

1. Each computer and each router connected to the Internet is given a unique address. The address is 32 bits long. It is expressed in what is known as dotted decimal format. For example, an address may be: 202.42.128.3. Each part of this four part address is decimal representation of a byte in the address (communication text books use the word *octet* instead of byte). Each country of the world has a clearing house to assign a range of addresses to Internet Service Providers who in turn give sets of addresses to organizations who are their subscribers.
2. Every computer and all routers connected to the Internet must use the IP protocol software. This software packetises data to be sent by a computer to another computer, attaches addresses and the serial number of the packet and facilitates transmission.
3. A router receives a packet, examines the destination address and sends it along the shortest route to the destination. If a route is busy, then it sends it by an alternate route if available. Traffic will be heavy if several computers attempt to send data simultaneously along a route. In such a case, a router will try to temporarily store some packets in the router and send them when a path is free. As the available storage in a router is limited the store may overflow leading to loss of packets. The Internet Protocol has no means of recovering lost packets. Thus, the Internet protocol is said to be "best-effort" packet delivery with no guarantee that all packets will be delivered.

3.4 TRANSMISSION CONTROL PROTOCOL

The "best effort" delivery of packets is not acceptable for reliable transmission of data. Thus another protocol called *Transmission Control Protocol*, TCP for short, is used along with Internet protocol (IP). The Transmission Control Protocol does the following:

1. We saw that packets may arrive out of sequence at a destination due to the fact that all packets do not take the same route. One of the jobs of TCP software is to examine the serial number of packets at the destination and reassemble them in the right sequence.
2. TCP software sends an acknowledgement to the source of a packet as soon as it is received. As packets may be lost due to congestion in the network, TCP software has a mechanism for the sender to retransmit a lost packet. TCP does it by estimating the time TE needed for a packet to reach its destination from the source and get back an acknowledgement. If no acknowledgement is received by the sender within TE, it retransmits the packet. The retransmitted packet is used to assemble the entire data. In some cases, a source may receive an acknowledgement after sending a duplicate packet. In such a case, the duplicate packet is discarded by the recipient. The time estimate TE would depend on the distance between the source and the destination and prevailing traffic in the network. TCP software automatically (adaptively) adjusts TE.

To summarize, TCP and IP work together to ensure reliable, economical transmission of data between any two computers connected to the Internet. Our discussions in the above two sections primarily describe the essentials of Internet. For interested students, detailed discussions may be found in Chapter 14 of Fundamentals of Computers, 4th edition by V. Rajaraman published by PHI Learning, New Delhi.

3.5 DOMAIN NAMES

The IP address in dotted decimal format is a string of digits which can be up to 12 digits long. It is difficult for people to remember such long strings of digits. Thus, a different addressing scheme using strings of characters is used to specify the identity of computers connected to the Internet. For example, the address of a server connected to the Internet is:

serc.iisc.ernet.in

In this address *serc* is the name of a host (computer) situated at the Supercomputer Education & Research Centre (SERC). This is called a subdomain of *iisc* which is the identity of Indian Institute of Science in which SERC is situated. IISc itself is a subdomain of *ernet* (Educational and Research Network) which is the Internet service provider's name. The topmost domain name is *in* indicating India, the country where the server is situated. Observe that the name is organized as a hierarchy.

All countries except USA use a 2-letter abbreviation for the country name as the top domain. For example, *uk* is for United Kingdom, *sg* for Singapore, *jp* for Japan, *de* for Germany (Deutch land). Some of the top level domain names in USA are given in Table 3.1. The country name is implied in the address only for USA.

Table 3.1 Top level domain names

Domain name	Use
com	Commercial organization
edu	Educational institute
gov	Government organization
mil	US military
net	Internet access provider
org	Non-profit organization

Every domain name can be translated into an IP address by a computer called a Domain Name Server. For example, if a user in USA wants to send a message to rajaram@serc.iisc.ernet.in, his or her request will be examined by a Domain Name Server in USA which will send a query to the Domain name server in India (as *in* is the top domain name). This machine will send a query to *ernet* name server which will in turn pass it on to the server at *iisc* which will have a table of IP addresses of all its departmental servers. The IISc mail server will now send the address query to the mail server *serc*. This server will have a table of names of persons in SERC and their current IP address. Searching this table gives IP address of *rajaram*. Thus, an Internet user need not know IP address but only the string of characters identifying a user. This is easy to remember. The secret of success of Internet technology is this decentralization of control. This decentralization has worked very well for over 20 years, but the recent explosion of the number of Internet users and the use of the Internet in e-commerce has led to a large number of disputes on the right to use a domain name which is easily identifiable by potential clients. For example, a company such as International Business Machines Corporation (IBM) would expect that it has the first right to use the domain name ibm.com which is easily recognized by customers as that of their company. If another firm Indian Basmati Merchants register a domain name in India ibm.co.in, International Business Machines will have a legitimate reason to object. Domain name disputes are common. In one case dispute on the rights to use a domain name arose between etoys.com, a toy store in USA and an artist group known as etoy who claimed the domain address etoy.com. There is a need to resolve such disputes. Assignment of domain names and IP addresses is controlled by an international authority known as Internet Corporation for Assigned Names and Number (ICANN).

This authority normally settles domain name disputes. We will revert to this later in this section. ICANN uses a hierarchical approach to decentralize assignment of IP addresses. For example, *in* (the top domain for India) represented by an authorized group in India is allocated a range of IP addresses. From this range *ernet* which is an Internet service provider is given a range of addresses. From this range *ernet* allocates a subset to *iisc* which in turn allocates a range of addresses to various departmental servers. The clients connected to a departmental server will be given unique IP addresses by the department only when they are connected to the Internet. This will allow dynamic allocation of a limited set of IP addresses. This is due to the fact that in the current Internet standard called IPv4, the number of addresses are limited as the IP address size is 32 bits long. With 32 bits it is possible to address only 4 billion addresses. These addresses are being quickly exhausted as the Internet is being used to even control appliances such as refrigerators and these appliances need IP address when connected to the Internet. A new TCP/IP standard called IPv6 is being introduced with 128-bit IP addresses which will provide an enormous increase in IP addresses.

Domain names are mapped to IP addresses but similar domain names may not have IP addresses close to one another. Thus, the domain names which are similar may be selected by groups in different parts of the world. The dispute between etoy and etoys arose because of this. There have been cases of individuals or groups registering a domain name similar to a well-known company and selling it to the legitimate party on payment. This is called *cybersquatting*. To resolve this problem ICANN has adopted a uniform domain name dispute resolution policy in October 1999. This policy states that an owner of a domain name being disputed by a complainant (who complains to ICANN) must obey the orders passed by an international cyberspace tribunal if accused of "bad faith" registration of a domain name (also known as cybersquatting). If a bad faith registration is proven, ICANN will remove the domain name of the squatter and assign it to the legitimate party. As the current system does not have a legal binding, a move is now on to refer international disputes to the world intellectual property organization which also resolves disputes on trademarks, copyrights, etc.

3.6 INTRANET AND EXTRANET

3.6.1 Intranet

We saw that all organizations nowadays have an organization wide LAN. Usually organizations are divided into departments such as Human Resources, Finance, Marketing, Purchase, etc. Each of these departments has their own LANs. Each individual normally has a desktop PC connected to the LAN. The LANs of each of the departments are interconnected with backbone hubs or switches depending on the number of LANs in the organization. An organization also centralizes and assigns functions such as institutional database, e-mail and heavy printing to servers.

Institutional networks may have diverse computers and operating systems. In order to ease communication among these machines it is desirable to use TCP/IP protocol. Several advantages accrue when this protocol is used. This protocol supports all Internet services, namely, e-mail, ftp, telnet, bulletin board and web services. All these application programs are available and no extra effort is needed to develop them. Thus, a company can have an e-mail service for communication among all employees. Notices can be posted on a company wide bulletin board. A company can implement a local web site giving information of relevance to the company's employees. Databases of different departments may be accessed by anyone provided he/she is authorized. For example, overtime of employees can be posted on the database of the accounts department by individual departments for use in computing the pay rolls. As another example, a book publisher can put in a web page the correct stock of all the books published and the up to date sales figures of all books region wise. This will help both the production department and the marketing department. In effect a lot of paper movement can be eliminated. Operations of the entire organization will be speeded up. A company wide computer network which uses TCP/IP protocol is called *Intranet*. Nowadays, all organizations have intranet as it provides all the Internet services mentioned above.

3.6.2 Extranet

No organization can work in isolation. It needs to communicate with many business partners. For example, an automobile manufacturing company requires frequent communications with its suppliers, distributors, servicing organizations and financial institutions. Communication with suppliers is essential to schedule production and to reduce inventories. Supply Chain Management (SCM) in which the manufacturer can track all suppliers to ensure that the inventory cost is minimized is one of the major applications of e-commerce.

For efficient supply chain management, the suppliers are allowed access to the production schedule and inventory status of the manufacturer on a need-to-know basis. This will be facilitated if the computers in the intranet of the supplier can have access to the appropriate information available in the automobile producer's intranet. This requires the intranets of the two organizations to be connected. As we pointed out in the last section, intranets use TCP/IP protocol. Thus, when the two intranets are connected, the expanded network will also use TCP/IP protocol. Such a connection of several intranets of cooperating businesses using TCP/IP protocol is called *extranet*. As all the cooperating businesses use TCP/IP protocol, they can automatically use all the facilities of the Internet such as e-mail, ftp, telnet and access internal web sites of each others business (based on what is allowed to be seen by trusted outsiders). This provides tremendous advantages to all the cooperating organizations to transact business.

The intranets of cooperating businesses can be interconnected either using the Internet which uses PSTN or using private leased lines from a communications service provider. Using PSTN is insecure as PSTNs are accessible to anyone. The communication between a manufacturer and its suppliers should be private as leakage of information to competitors will be detrimental to both suppliers and manufacturers. Thus, there is a need to use a secure connection. Private leased lines are secure. They are, however, expensive. There is another method of using the Internet with enhanced security which is called *Virtual Private Network* (VPN). In a VPN security of transactions is ensured using a protocol called *Internet Protocol Security Architecture* (IPSEC). This protocol adds an additional security layer to TCP/IP protocol. This layer is created by encapsulating an IP packet in a new secure IP packet. This encapsulation is performed by IPSEC compliant routers at the boundary of each intranet or by IPSEC compliant server in the intranet which forwards IP packets to the Internet. The encapsulation process encrypts the data within the IP packet as well as IP address using a secret key. (Encryption replaces a bit string by another string by transforming the bit string using a key. We will describe encryption in detail later in this book). The IPSEC packet format is shown in Figure 3.3.

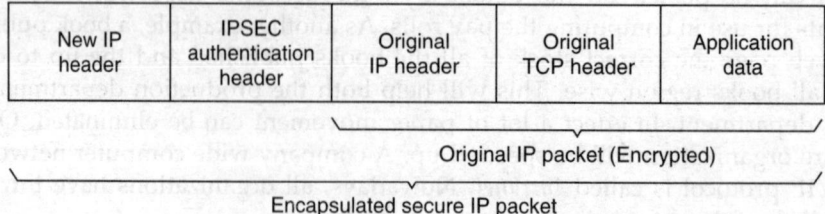

Figure 3.3 IPSEC packet format.

The encapsulated secure packet can be decoded only by IPSEC compliant routers and servers having access to the secret encryption key. Messages sent using IPSEC protocol on the Internet between two IPSEC compliant routers/servers is known as *tunneling*. Thus, intranets of businesses may be connected together using the Internet employing IPSEC compliant routers at their boundaries where they are connected to the Internet as shown in Figure 3.4. Several Internet Service Providers implement VPN between cooperating businesses by using the Internet. VPN is thus cheaper than leased lines to create extranets.

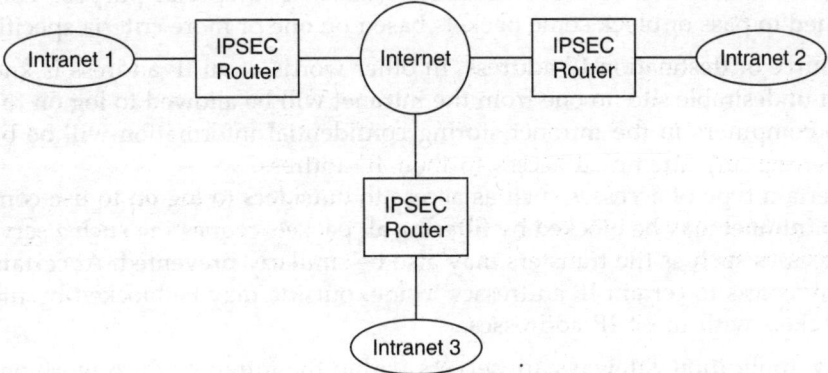

Figure 3.4 VPN implementation.

3.7 FIREWALLS

A *firewall* is a combination of software and hardware, which protects an organization's intranet from mischievous or unwelcome intrusion by users of the Internet. A firewall is also meant to prevent the users of an organization's intranet from accessing Internet sites considered undesirable by the managers of the organization. For example, a company management would not like its employees to access gaming sites or sites with known security threats.

We saw that the intranets of organizations are connected to the Internet using routers. The router connects the organization to its Internet service provider (ISP). ISP may have a large number of clients. An organization has no control over who sends messages to it and will not know whether it is a useful message or not. If it is an e-commerce site, it should be accessible to anyone. Also organizations maintain web-presence to let anyone access the site to get information. The main point is that the Internet is "open" and allows anyone to access it. Thus, there may be undesirable persons who may like to disrupt an organization's functioning by accessing computers in it. Thus, access to computers must be controlled.

The types of undesirable intrusions which have been observed are:

- Attempts to access secret information such as credit card numbers, sales and client information, valuable drawings, etc., stored in an organization's databases.
- Erasing or changing information on a web page. In general, mutilating the web page of an organization.
- Attempts to prevent legitimate users from accessing an organization's intranet by monopolizing a resource. Some examples are flooding it with e-mail, sending forged TCP connection-establishment segments to a host which fills up TCP buffer space preventing legitimate users from accessing the host.

Thus, the main purpose of a firewall is to prevent undesirable intrusions discussed above.

There are two main methods used to implement the functions of a firewall. One of them which is simple and not expensive is to use *packet-filtering*. The other more complex firewall uses a computer at the boundary of the intranet and is called a *proxy application gateway*. We will now describe the functions of each of these.

Packet filtering is normally implemented using the router which connects an organization's intranet to its ISP. The router (which is a special purpose computer) is programmed to pass or block some packets based on one or more criteria specified below:

- Source or destination IP address. In other words, if an IP address is known to be an undesirable site no one from the intranet will be allowed to log on to it. Access to computers in the intranet storing confidential information will be blocked by filtering any attempted access to their IP address.
- Certain type of accesses such as access to outsiders to log on to use computers in an intranet may be blocked by filtering all packets requesting such a service. Other accesses such as file transfers may also be similarly prevented. At certain times of day access to certain IP addresses inside/outside may be blocked by filtering out packets with those IP addresses.

Proxy application gateways are servers within the intranet which work on behalf of user(s) by performing certain specified functions. These are more complex than packet filtering. Some examples are:

- Filter data accessed with certain keywords, for example, any data which is marked company confidential.
- Check for viruses in data files entering an intranet.
- Prevent access to some applications with known security holes.
- Create log files and audit trails of access by users to certain sites, files, and time at which they were accessed and time spent in such activities.
- Provide network address translation (abbreviated NAT) which converts internal IP addresses used within the intranet to those recognized by (or registered with) the Internet. The intranet's network may be very large with local IP addresses which are not all known(or registered) with the Internet. Only a subset of addresses may be used for transaction with computers outside the organization. By this address translation method outsiders will not know the IP addresses used within the organization by various computers. This is a good method of ensuring security.

Firewall using proxy application gateway replace the source address of the transaction with its own IP address. Thus, outsiders will only know this address and will not be able to access any other computer in the intranet.

Besides the two types of firewalls we have described, there is one more firewall known as application firewall. This is a software implemented in a computer at the gateway of an intranet which scans for viruses in the incoming and outgoing files. *Virus* is a malicious program which infects other programs by modifying them and puts a copy of itself so that when this program is sent to another computer it is infected with the virus. It also filters unsolicited e-mails (known as *spam*) and blocks programs entering the intranet which are suspected to be hiding malicious codes. Some malicious codes may seem useful but will be hiding spy programs to capture passwords or steal data. These are known as *Trojans*. Malicious codes known as *worms* are similar to viruses but propagate from one computer connected to a network to another automatically using the network to spread. In Figure 3.5 we give a block diagram of various firewalls and their connections between the Internet and intranets.

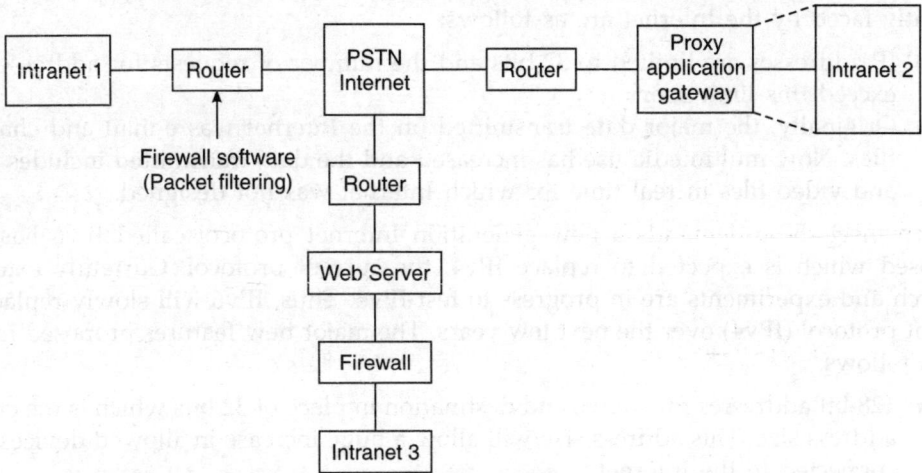

Figure 3.5 Various types of firewalls and their connections to intranets.

We have seen that TCP/IP protocol is almost universally used by the Internet. In some cases where packet loss may be tolerated, another protocol called UDP (*user datagram protocol*) is used instead of TCP. This is called a connectionless protocol as datagrams may be received in any order (and not serialized) and there is no acknowledgement from the receiver back to the sender as in TCP. Thus, lost packets are not resent as the sender assumes that all packets reach safely. By default, firewalls filter out all UDP packets as it is not possible to uniquely identify the sender of the packet. Thus, special action should be taken to modify firewalls if an organization wants to allow UDP packets into the intranet and send them out to the ISP.

3.8 THE FUTURE OF INTERNET TECHNOLOGY

From humble beginning in the late 60s, today the Internet spans the whole world with 1.6 billion of computers connected to it. The remarkable achievement of the Internet technology is that it has been able to accommodate exponential growth (i.e., doubling each year) of the number of computers connected to it. Also the speed of computers has been doubling every year and there are a variety of computers connected to the Internet. Individual LANs in an organization may use any type of interconnection and any local protocol for communication. The physical connection between computers may range from fast gigabit fibre optics to slower wireless. In spite of this variety of technologies and speeds, the Internet still works effectively. The reason for this is the universal adoption of TCP/IP as the standard protocol which has proved very robust in spite of rapid changes in technology. TCP/IP protocol emerged as a result of cooperative effort in which a large number of organizations participated and experimented before accepting any version. The Internet protocol accommodates a variety of hardware and a variety of network speeds as it makes no assumptions regarding the underlying network hardware. Packet switching ensures efficient and fault tolerant routings of packets. TCP ensures reliable receipt of all packets sent by a sender to a receiver. It continuously monitors traffic conditions on the Internet and automatically adapts when there is congestion in the network. The only major problems currently faced by the Internet are as follows:

- IP addresses are limited to 32 bits and the number of requests for addresses will exceed this limit soon.
- Originally, the major data transmitted on the Internet was e-mail and character files. Now multimedia use has increased and the data transmitted includes audio and video files in real time for which Internet was not designed.

To meet these demands a new generation Internet protocol called IPv6 has been proposed which is expected to replace IPv4, the current protocol. Currently extensive research and experiments are in progress to test IPv6. Thus, IPv6 will slowly replace the current protocol (IPv4) over the next few years. The major new features proposed in IPv6 are as follows:

- 128-bit addresses for source and destination in place of 32 bits which is the current address size. This address size will allow a huge increase in allowed devices to be connected to the Internet.
- The packet lengths can go up to 64 KB. This will allow easy transfer of multimedia data, particularly, voice.
- The protocol has a feature to specify the kind of data being transmitted and will permit real-time multimedia data to be given priority in transmission.
- The proposed protocol has in-built security of data being sent over the Internet.

It is thus clear that the Internet has come to stay and will continue to grow in the coming years. With increasing use of wireless and mobile systems, many ordinary household systems such as refrigerators, ovens, etc., are being connected to the Internet and controlled remotely. Thus, the growth and utility of Internet will be enormous.

SUMMARY

1. Computers in an organization are normally connected together by what is known as a Local Area Network (LAN). This allows computers to communicate with one another and share resources.
2. Normally in a LAN certain computers provide services such as database storage, printing, etc., and are called servers. Other computers which use these services are known as clients.
3. For computers made by different manufacturers to communicate and cooperate in solving problems, it is necessary to implement a set of agreed upon rules for exchanging messages. This is called a communication protocol.
4. Computers connected to LANs in two different organizations communicate via a special purpose computer called a router.
5. Using routers, LANs in different geographical locations can be interconnected. Such an (international) network of interconnected LANs is called the Internet.

6. A router can route data from a computer in a LAN to another one in another LAN only if it knows the address of the intended receiver. A system of assigning addresses to each computer connected to the Internet has been standardized.
7. The address of a computer connected to the Internet is a 32-bit address. It is expressed in what is known as a dotted decimal format. An example address is 202.42.128.3. Each part is a decimal representation of a byte. This is known as the IP address of the computer.
8. Each country in the world has a clearing house to assign a range of IP addresses to Internet service providers (ISPs) which in turn give sets of addresses to their subscribers.
9. The method of communicating data using the Internet employs a set of rules called the Internet Protocol abbreviated as IP. In this protocol messages are divided into fixed size packets (called user datagrams). The packets are sent along available routes from a source computer to a destination computer.
10. IP protocol does not guarantee delivery of all packets. Packets may also arrive out of order. This problem is solved by a protocol called Transmission Control Protocol (TCP). This protocol puts received packets in the correct serial order and arranges retransmission of lost packets.
11. All computers connected to the Internet and all routers use TCP/IP protocol software for reliable packet transmission.
12. An IP address is converted into a string of characters called a domain name system. The domain name system is easy to use. It is controlled by an international authority known as Internet Corporation for Assigned Names and Numbers (ICANN).
13. LAN of an organization which uses the TCP/IP protocol for communication is called an intranet.
14. Intranets of cooperating organizations connected using communication links is known as an extranet.
15. Intranets can be connected using PSTN or leased lines. Leased lines are expensive. PSTNs are insecure. Thus encrypted IP packets encapsulated in another packet are sent using PSTN. This is called Virtual Private Network.
16. A firewall is a combination of software and hardware, which protects an organization's intranet from undesirable intrusion.
17. There are two types of firewalls—one is called a packet filtering firewall and is implemented in the router to pass or block some packets based on specified criteria, and the other is called a proxy application gateway which is a computer connected to an organization's intranet for performing a variety of tasks such as virus detection, filtering data based on content, IP address hiding, etc.

EXERCISES

3.1 Define the Internet. Why is it important in e-commerce?
3.2 What is a protocol? What is the protocol used by the Internet?
3.3 What is packet switching? Does the Internet use packet switching? If yes, why?

3.4 What is a router? Where is it used?
3.5 What is the size of a packet used in the Internet?
3.6 Give the format of a packet. What is the maximum number of bytes in a packet?
3.7 What is Internet Protocol (IP)? Describe it.
3.8 Why is Internet Protocol called "best effort"?
3.9 What is IP address? How many bytes does it have?
3.10 What is dotted decimal format for IP address? Why is it used to represent IP address? What is the maximum number of IP addresses in the current Internet standard?
3.11 What is TCP? Describe the function of TCP.
3.12 Why is TCP required in the Internet?
3.13 What is a domain name? Where is it used?
3.14 Explain how a domain name system will find the IP address of an e-mail address skb@ipc.iisc.ernet.in
3.15 What is cybersquatting? How is the right to use a domain name determined?
3.16 What is the expansion of ICANN? What is the primary responsibility of ICANN?
3.17 What is an intranet? What is the protocol used by intranet?
3.18 What are the advantages which accrue to an organization if they implement an intranet?
3.19 What is an extranet? What advantages accrue to organizations connected to an extranet?
3.20 Explain how an extranet could be useful in supply chain management.
3.21 What is VPN? Where is it used?
3.22 Explain briefly how VPN is implemented to ensure security.
3.23 What is a firewall? Where is it used? Why is it necessary?
3.24 What is packet filtering firewall? Explain its working.
3.25 What is a proxy application gateway? What functions does it perform?
3.26 What is IPv6? What are the differences between IPv4 and IPv6?

OBJECTIVE QUESTIONS

Each question has four possible answers. Pick the most appropriate answer.

3.1 The Internet is
   (a) A local computer network
   (b) A worldwide network of computer networks
   (c) An interconnected network of computer networks
   (d) A worldwide interconnected network of computer networks which uses a common protocol to communicate with one another
3.2 The facilities available in the Internet are
   (i) Electronic mail
   (ii) Remote login

(iii) File transfer
  (iv) Word processing
    (a) (i), (ii)   (b) (i), (ii), (iii)
    (c) (i), (ii), (iv)   (d) (ii), (iii), (iv)

3.3 The Internet requires
  (a) An international agreement to connect computers
  (b) A local area network
  (c) A commonly agreed upon set of rules to communicate between computers
  (d) A World Wide Web

3.4 Each computer connected to the Internet must
  (a) Be an IBM PC   (b) Have a unique IP address
  (c) Be Internet compatible   (d) Have a modem connection

3.5 IP address is currently
  (a) 4 bytes long   (b) Available in plenty
  (c) 6 bytes long   (d) Not assigned as it is all used up

3.6 IP addresses are converted to a
  (a) Binary string   (b) Dotted decimal string
  (c) Hierarchy of domain names   (d) Hexadecimal string

3.7 Internet addresses of individuals must always have at least
  (i) A country name   (ii) Internet service provider's name
  (iii) Name of an organization   (iv) Name of an individual
  (v) Type of organization
    (a) (i), (ii), (iii)   (b) (ii), (iii), (iv)
    (c) (i), (iii)   (d) (iii), (iv), (v)

3.8 The Internet uses
  (a) Packet switching   (b) Circuit switching
  (c) Telephone switching   (d) Telex switching

3.9 Internet data is broken up as
  (a) Fixed length packets   (b) Variable length packets
  (c) Not packetized   (d) 64-byte packets

3.10 An Internet packet data structure consists of
  (i) Source address
  (ii) Destination address
  (iii) Serial number of packet
  (iv) Message bytes
  (v) Control bits for error checking
  (vi) Path identification bits
    (a) (i), (ii), (iii)   (b) (i), (ii), (iii), (iv)
    (c) (i), (ii), (iii), (iv), (v)   (d) (i), (ii), (iii), (iv), (v), (vi)

3.11 The packets of an Internet message
  (a) Take a predetermined path
  (b) Take a path based on packet priority
  (c) Go along different paths based on path availability
  (d) Take the shortest path from source to destination

3.12 The time taken by Internet packets
   (a) Can be predetermined before transmission
   (b) Are same for all packets
   (c) May be different for different packets
   (d) Is irrelevant for audio packets

3.13 Internet packets
   (a) May be lost in transit
   (b) Are always delivered without fail
   (c) Are resent if lost in transit
   (d) Are received in the same order as they were sent

3.14 The receiver of an Internet packet
   (a) Does not acknowledge receipt of packets
   (b) Always sends an acknowledgement of receipt to the sender
   (c) Sends no acknowledgement only when the traffic is light
   (d) Sends acknowledgement only if the network is congested

3.15 TCP stands for
   (a) Transport Control of Packets
   (b) Terminal Computer Policy
   (c) Transmission Control Protocol
   (d) Transit Communication of Packets

3.16 The main purpose of TCP is to
   (a) Make sure that all packets sent are received
   (b) Ensure that all packets belonging to a message take the same route
   (c) Ensure that all packets are received and in the correct order
   (d) Control of all transmissions of packets

3.17 By an intranet, we mean a
   (a) LAN of an organization
   (b) Wide Area Network connecting all branches of an organization
   (c) Corporate computer network
   (d) Network connecting all computers of an organization using TCP/IP protocol

3.18 By an extranet, we mean
   (a) An extraordinarily fast computer network
   (b) The intranets of cooperating organizations interconnected via a secure connection
   (c) An extra network used by an organization for higher reliability
   (d) An extra connection to Internet provided to cooperating organizations

3.19 Advantages of the intranet are
   (i) All internet applications such as ftp and email can be used within the organization
   (ii) Diverse computers in an organization can be interconnected to create a unified system
   (iii) Client-server architecture can be implemented
   (iv) It uses TCP/IP protocol
     (a) (i), (ii)   (b) (i), (iii)
     (c) (iii), (iv)   (d) (ii), (iii)

3.20 Extranet permits
   (a) Logging into any computers in the network by any user
   (b) Accessing database of an organization from another organization if permitted
   (c) Secure transmission of packets
   (d) Extra length IP packets

3.21 A firewall is a
   (a) Wall built to prevent fires from damaging a corporate intranet
   (b) Security device deployed at the boundary of a company to prevent unauthorized physical access
   (c) Security device deployed at the boundary of a corporate intranet to protect it from unauthorized access
   (d) Device to prevent all accesses from the Internet to the corporate intranet.

3.22 A firewall may be implemented in
   (a) Routers which connect intranet to the Internet
   (b) Bridges used in an intranet
   (c) Expensive modem
   (d) User's application programs

3.23 A firewall as part of a router program
   (a) Filters only packets coming from the Internet to an intranet
   (b) Filters only packets going to the Internet from an intranet
   (c) Filters packets travelling from and to the intranet from the Internet
   (d) Ensures rapid traffic of packets for speedy e-commerce

3.24 Filtering of packets by firewalls based on a router has facilities to
   (i) Prevent access to the Internet to some clients in the intranet
   (ii) Prevent access to Internet at certain specified times to certain clients
   (iii) Filter packets based on source or destination IP address
   (iv) Prevent access by certain users of the Internet to other specified users of the Internet
      (a) (i), (iii) (b) (i), (ii), (iii)
      (c) (i), (ii), (iii), (iv) (d) (ii), (iii), (iv)

3.25 One function of a proxy application gateway firewall is
   (a) Allow corporate users to use efficiently all Internet services
   (b) Allow intranet users to securely use specified Internet services
   (c) Allow corporate users to use all Internet services
   (d) Prevent corporate users from using Internet services

3.26 Proxy application gateway
   (i) Acts on behalf of all intranet users wanting to access the Internet securely
   (ii) Monitors all accesses to the Internet and allows access to only designated IP addresses
   (iii) Disallows use of certain protocols with security problems
   (iv) Disallows all Internet users from accessing an intranet
      (a) (i), (ii) (b) (i), (ii), (iii)
      (c) (i), (ii), (iii), (iv) (d) (ii), (iii), (iv)

3.27 Extranet normally connects intranets using
   (a) PSTN
   (b) Internet
   (c) Leased lines
   (d) Modems

3.28 A Virtual Private Network
   (a) Is used to securely interconnect intranets to form an extranet
   (b) Uses a private leased network to interconnect intranets
   (c) Uses Internet to connect intranets
   (d) Is a secure extranet

3.29 IPSEC
   (a) Is a secure IP packet
   (b) Adds an additional security layer to TCP/IP protocol
   (c) Encrypts data in an IP packet using a secret key and appends a new IP address to the packet
   (d) Is used in intranet

3.30 Messages sent using IPSEC
   (a) Use a secure net
   (b) Use leased line connection
   (c) Are between routers of two intranets
   (d) Are between IPSEC compliant routers/servers

3.31 A virus is a malicious program which
   (a) Propagates using the Internet
   (b) Infects other programs with whom it comes into contact
   (c) Infects another program by modifying it and putting a copy of itself in it to enable spreading
   (d) Spreads through floppy disks

3.32 A Trojan is a malicious program which
   (a) Looks like a virus
   (b) Propagates via e-mails
   (c) Masquerades as a useful program but is a security threat
   (d) Has the appearance of a useful code but it hides spy programs which captures confidential data and sends them out

3.33 A worm is a malicious program which
   (a) Is similar to a virus but spreads to other computers connected to the network automatically
   (b) Is a variant of a virus program
   (c) Worms itself via network connection
   (d) Is a mutation of a virus

CHAPTER 4

Network Services

LEARNING GOALS

In this chapter, we will learn:
1. About various services available using the Internet infrastructure.
2. About the World Wide Web and its importance in e-commerce.
3. Basics of design of a web page.
4. Search engines and their applications.

4.1 INTRODUCTION

In the previous chapters, we described the physical layer and the logical layer of e-commerce layered architecture proposed by us. These two layers are the foundations over which the next layer, namely, network services are created. The primary network services essential for e-commerce are the *World Wide Web* and *browsers* used to navigate the web. The primary purpose of this chapter is to explain how they work. One of the main purposes of navigating the web is to search for information. Ability to search for information on dealers of various products, e-shops, hotels, airline schedules, etc., is the main reason why web became valuable to all businesses and individuals. We will explain how search engines are used to retrieve relevant information from the World Wide Web. We will follow this up by briefly describing how web pages are designed using a language called HyperText Markup

Language (HTML). Lastly, browsers provide several other services such as e-mail, telnet and remote file transfer. We will describe the basics of these services.

4.2 THE WORLD WIDE WEB

The idea of the World Wide Web originated in the high energy particle physics laboratory where the physicist Tim Berners-Lee was trying to find a convenient method of exchanging documents among researchers in particle physics, spread all over the world. This was in 1989 when the Internet was becoming a reliable information exchange infrastructure. His purpose was to build a software which would use this infrastructure effectively to retrieve documents stored in various servers spread across laboratories. He designed a new set of rules for computers to retrieve textual data from remote servers. This was called *HyperText Transfer Protocol* (HTTP). The next step was to describe a notation to mark keywords in documents for ease of search and retrieval. This was called *HyperText Markup Language* (HTML). The final step was to assign a unique address specifying the server connected to the Internet in which the document resided and how to access it. It was called a *Universal Resource Locator* (URL). These were the basic steps necessary to facilitate easy retrieval of desired documents from a distributed interlinked set of computers.

These steps and the relevant software were essential but not sufficient to allow easy retrieval of documents on a variety of subjects. The next crucial step was the development of *hypertext browser* by Marc Andersen and his team of programmers at the National Centre for Supercomputer Applications at the University of Illinois. This was called Mosaic Web Browser and was the software which really opened up the World Wide Web as we know it today. Mosaic provided a good graphical user interface and made the web accessible to the general public who do not have any special programming knowledge. The browser technology has continuously improved, and the currently popular web browsers are the Internet Explorer designed by Microsoft and integrated with Windows OS, Netscape Communicator which is the successor of Mosaic and Mozilla Firefox. These browsers are freely available and easy to use.

4.2.1 Hypertext

Assume a text such as that in Figure 4.1 is given. This text gives a brief write-up about Bangalore in which a number of keywords are shown in a different font. These keywords are linked to other pages which give information in more detail regarding the keywords.

About Bangalore

Bangalore is the capital of Karnataka. It is called the Silicon Valley of India as it has a large number of InfoTech industries. Leading among them are INFOSYS, WIPRO, TCS and SONATA. It is also the home of the Indian Institute of Science, a pioneering educational and research institution which was founded in 1907 by Sir J.N. Tata. It also has many engineering colleges. It is also called the garden city as it has one of the largest and oldest botanical gardens of India called Lalbagh. Lalbagh has several flowering trees and some rare birds with wonderful chirping sounds.

Figure 4.1 A document with keywords linking them to other documents (keywords are shown in a different font).

This idea of embedding selectable keyword in a text, which links it to other documents, is called *hypertext*. The hypertext facilitates what may be called nonlinear reading, i.e., while reading a document stored in a computer if some topic interests you, it is possible for you to click on the keyword corresponding to that topic. After reading the document, you could return to the original document to the same point where you left it. In the computer's memory (normally, disk) the documents are linked as shown in Figure 4.2. This figure is given just to satisfy your curiosity.

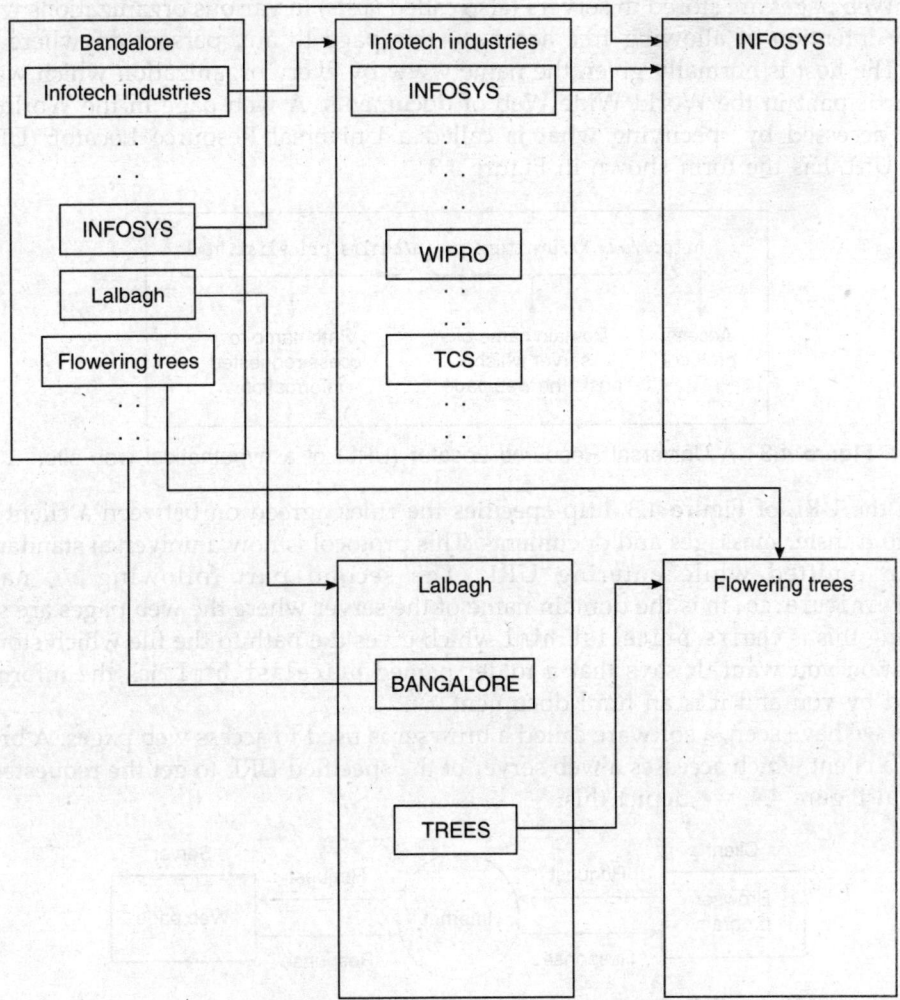

Figure 4.2 Hypertext links between documents.

Given an ordinary text, a language to format the document and mark the key words in the text is called hypertext markup language. This language is now popular to create documents called *web pages* which are linkable to other web pages on the World Wide Web.

4.2.2 Universal Resource Locator

We stated in the last section that a hypertext is organized by specifying a number of keywords and the links from these keywords to other relevant documents. The major innovation in World Wide Web browser is that it specifies the address of the server where the document, also called a web page, is kept, and this address is displayed.

The World Wide Web is organized using web pages. Web pages use hypertext with a set of clickable (i.e., selectable) keywords which are normally highlighted in a different colour. Web pages are stored in servers (also called *hosts*) in various organizations with the primary intention of allowing free access to this page to any person, anywhere in the world. The host is normally given the name www by every organization which wants to be a participant in the World Wide Web of documents. A web page in the World Wide Web is accessed by specifying what is called a Universal Resource Locator (URL). A typical URL has the form shown in Figure 4.3.

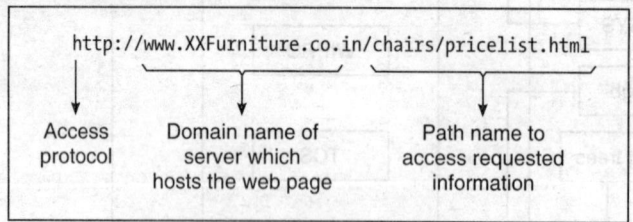

Figure 4.3 A Universal Resource Locator (URL) of a hypothetical web site.

In the URL of Figure 4.3, http specifies the rules agreed on between a client and a server to transfer messages and documents. This protocol is now a universal standard and is often omitted while entering URL. The second part following ://, namely, www.XXFurniture.co.in is the domain name of the server where the web pages are stored. Following this is chairs/pricelist.html which gives the path to the file which stores the information you want. It says that a folder named pricelist.html has the information required by you and it is an html document.

As we have seen, a software called a browser is used to access web pages. A browser runs in a client which accesses a web server of the specified URL to get the requested web pages. In Figure 4.4, we depict this.

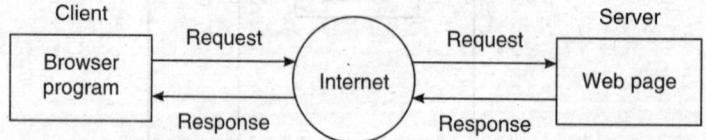

Figure 4.4 Browser and web server as a client-server system.

A request from a browser travels via the Internet to the web server. The server responds and returns the requested web page. A sample page is shown as Figure 4.5. In many cases, you may not know the explicit path to a specific file. Usually, all sites have a top page called *home page* which contains a table of contents. You can click on any of the items you are interested in and the browser will automatically go to the relevant document, display it and also show the URL in the address field of the page This is illustrated in

Figure 4.6. If you click on catalogue on this page, it will take you to the page with lists of all types of items sold. You can click on the specific item type, e.g., chairs, you are interested in and it will display information on that item.

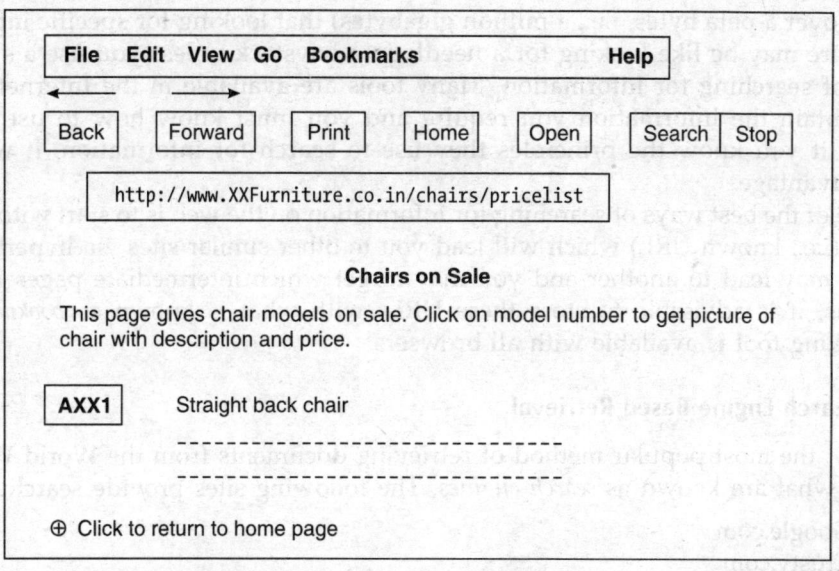

Figure 4.5 Display of requested URL.

(We have shown a typical screen you will see. It is not that of any particular browser).

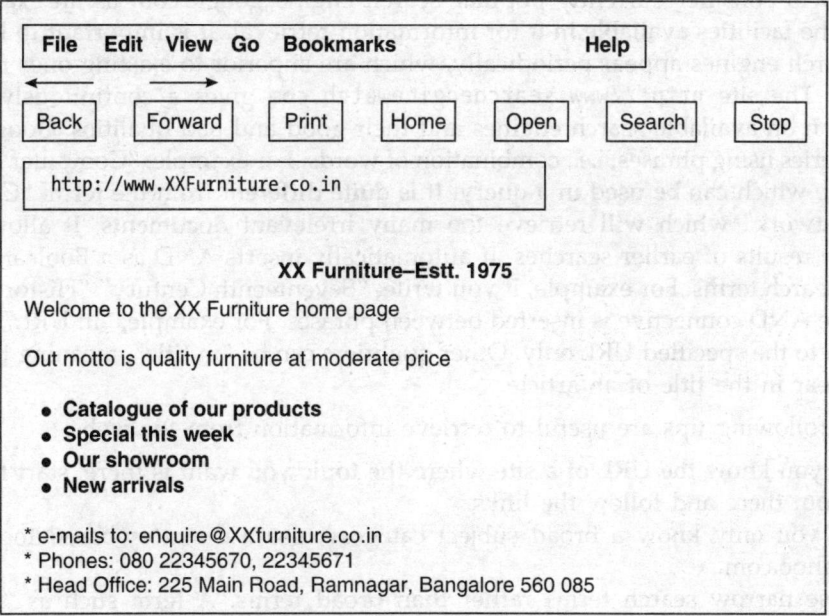

Figure 4.6 Display of main web page of XX furniture. Any of the items in bold face may be clicked to go to the relevant page.

4.3 INFORMATION RETRIEVAL FROM THE WORLD WIDE WEB

As of today, there are over 200 million hosts on the World Wide Web serving over a billion users all over the world. The web is thus a vast mine of useful information. It is so huge (over a peta bytes, i.e., a million gigabytes) that looking for specific information you require may be like looking for a needle in a haystack unless you use a systematic method of searching for information. Many tools are available in the Internet to assist you to obtain the information you require and you must know how to use them. In addition, if you know the principles they use to search for information, it will be an added advantage.

One of the best ways of searching for information on the web is to start with a known web site (i.e., known URL) which will lead you to other similar sites via hypertext links. One link may lead to another and you may forget which intermediate pages interested you. Thus, it is advisable to store these URLs with what is known as *bookmarks*. The bookmarking tool is available with all browsers.

4.3.1 Search Engine Based Retrieval

Currently, the most popular method of retrieving documents from the World Wide Web is to use what are known as *search engines*. The following sites provide search engines:

- Google.com
- Clusty.com
- Yahoo.com
- Bing.com

We will consider currently popular search engine google.com as an example and examine the facilities available in it for information retrieval. It is important to know that newer search engines appear periodically, which are superior to existing ones for certain purposes. The site http://www.searchenginewatch.com gives a continuously updated information on available search engines and their good and bad qualities. Google allows search queries using phrases, i.e., combination of words. For example, "Computer Network" is a phrase which can be used in a query. It is quite different from the term: "Computer" *AND* "Network" which will retrieve too many irrelevant documents. It allows search within the results of earlier searches. It automatically inserts AND as a Boolean operator between search terms. For example, if you write: "Seventeenth Century" "History" "South India", the AND connective is inserted between phrases. For example, "in URL" will limit the search to the specified URL only. Other qualifiers can be "in title", meaning the phrase must appear in the title of an article.

The following tips are useful to retrieve information from the web.

- If you know the URL of a site where the topic you want is there, start the search from there and follow the links.
- If you only know a broad subject category, use a directory-based tool such as yahoo.com.
- Use narrow search terms rather than broad terms. A term such as "computer network" will retrieve too many documents. If you are interested in *local area network design*, use it as a search phrase.

- Use specific and somewhat unusual words in your query which will narrow down the search. General terms will flood you with documents. Use as many synonyms as you know of the keyword in your search query.
- Some search tools are sensitive to word order and take the first word as most important. So use the most important keyword first in your query.
- If the search allows NOT operator, use it to exclude information you do not want.
- Some search engines rank order retrieved documents, giving the most relevant ones first and irrelevant ones later. This may not always be the correct order. So see at least the first four or five pages retrieved to choose what you want.
- Create bookmarks of useful sites so that you can use them later if you need them.

Before we conclude this section, we will briefly explain how search engines function. Search engines such as google do not use people to index documents. Documents are automatically indexed by *web crawler* programs. These programs systematically visit web pages and pick those containing the specified keyword(s) and create an index. Web crawling can get out of control. So crawlers use some method of limiting the number of branches taken. They also rank web pages by their importance and relevance and give them as first few items when you search. Importance is normally determined by how many URLs refer to the web page in their hypertext link, the number of words in the web page which match those used in the search query, the number of times the web page has been referred to, etc. Design of search engines and relevance ranking has been a research area for many years. Continuous improvements are taking place based on experimentation and in a few years excellent search tools are bound to appear. This will make it easy for lay users to retrieve all the relevant documents satisfying their query from the World Wide Web.

4.4 MARKUP LANGUAGES

By *markup* we mean a symbol added to a document which adds special meaning to the document. A markup language is used to describe the content and structure of a document. A markup language called Standard Generalized Markup Language (SGML) is an international open standard which is independent of a computer's hardware or operating system. SGML description of a document consists of three parts:

1. *Declaration:* This is a header file that contains system-specific information that is needed to enable the document to be used and modified by the system on which it is stored and viewed.
2. *Document Type Definition (DTD):* This is a hierarchical tree-structure definition for the document elements. Examples of document elements are: title, heading, paragraphs, and so on. *Tags* assign element names to the content of the document. For example:

    <DAY> Friday </DAY>

    <TITLE> Displaying Textual Data </TITLE>

    Here the Tags are <DAY> and <TITLE>.
    DTD has been extensively used to describe documents.

3. *The document instance*: This is the actual document with added tags in the correct order as determined by SGML specification.

Even though SGML is the original attempt to formalize the description of documents, other languages based on it have become more popular and widely used. They are HTML and XML. HTML is primarily used to describe documents stored as web pages on the World Wide Web. XML is essential to exchange documents such as purchase order and invoice in electronic form among cooperating businesses.

4.4.1 Hypertext Markup Language (HTML)

This is a language used to describe pages stored on the World Wide Web. It is used to create a visually appealing presentation of a document to a reader, incorporate graphics, links to other parts of the web document and web pages stored in other sites. When pages using HTML are stored in the World Wide Web server, standard Browser software such as Netscape and Internet Explorer can be used to view the documents and also other documents linked to it. You may wonder why we should know about HTML. The main reasons are to understand how hypermedia is created and for you to appreciate the amount of effort needed to create a web page.

Figure 4.7 gives an idea about the way in which an HTML document is described. This document will be displayed as shown in Figure 4.8. In the HTML description, symbols <HTML> and </HTML> enclosed in angular bracket are called tags. They surround the other tags in the description of the document and tell the browser that it is an HTML page. <TITLE> is used to identify the document in the web browser's title bar and is stored as the *bookmark* of this document. <HEAD> and </HEAD> mark the boundaries of TITLE information. The contents displayed to a user who browses the document is that within page. The HTML description is quite comprehensive and books have been written describing it. In this brief introduction, we will only give an idea of how HTML is used to describe

```
<HTML>
<HEAD>
<TITLE> Book on E-commerce </TITLE>
</HEAD>
<BODY BGCOLOUR = "WHITE", TEXT = "BLACK">
<H1> Essentials of e-commerce technology</H1>
<H2> A first level book on e-commerce </H2>
<P> Publisher: <I> PHI Learning Pvt.Ltd.</I></P>
<P><CENTRE><B> UNIQUE FEATURES </B></CENTRE></P>
<UL>
        <L1> Concise, lucid, presentation
        <L2> Emphasis on technology
        <L3> Multiple choice questions for self-learning
</UL>
</BODY>
</HTML>
```

Figure 4.7 HTML description of a document.

<BODY> and </BODY>. The body tag specifies various colours to be used to display the documents. Those interested may look at http://www.w3.org/pub/www/ published by the World Wide Web consortium. The tag <H1> specifies the section heading in bold face, larger size font. There are six levels of headers <H1> to <H6> with decreasing size of letters in headings. <P> indicates the beginning of a new para. Observe that the beginning of a description by a tag such as <P> should end with a corresponding tag </P>, where a forward slash (/) is used in front of the letter. The other tags we have used are: <I> indicating italics and <B> indicating bold face. The tag <CENTRE> is to centre the line. Observe the tag <UL> in the HTML document of Figure 4.7. This denotes an unordered list with each item marked with a bullet. An ordered list has tag <OL> which enumerates each item on the list. If we write a HTML description of an ordered list (See Figure 4.9), the browser will number and list items as shown in Figure 4.10.

ESSENTIALS OF E-COMMERCE TECHNOLOGY

A first level book on e-commerce

Publisher : *PHI Learning Pvt.Ltd.*

UNIQUE FEATURES

- Concise lucid presentation
- Emphasis on technology
- Multiple choice questions for self-learning

Figure 4.8 Display of document described by HTML of Figure 4.7 by a standard browser.

```
<OL>
<L1> Introduction
<L2> Earlier work
<L3> Contributions
<L4> Conclusions
</OL>
```

Figure 4.9 An ordered list.

1. Introduction
2. Earlier work
3. Contributions
4. Conclusions

Figure 4.10 An enumerated list displayed by browser when seen on a browser.

The above HTML description is for a very simple document. Facilities are available in HTML to display images described as either tif or jpeg file in the document. The image file is specified by the tag IMG. For example, <IMG src = "mypicture.tif">. Observe that there is no end tag in this case. It is called a standalone tag. The letters src following IMG indicate the source of the image file. It is also possible to display tables and forms. In fact, many of the facilities available in word processor to design a visually pleasing document are also available in the latest version of HTML.

Hyperlinks in HTML are links to other web pages from a given document. Such links are created using a tag called *anchor tag* represented by <A>. Such an anchored link is included in an HTML document as shown in Figure 4.11.

Bangalore is the home of
<A href = "http://www.iisc.ernet.in> Indian Institute of Science </A>

Figure 4.11 Linking using anchor tag.

This is displayed by the browser with the anchored text underlined and in a different colour as shown below:

Bangalore is the home of <u>Indian Institute of Science</u>.

When the underlined text is clicked the browser loads the web site of Indian Institute of Science specified in the HTML document.

It is always not necessary to write HTML description for many documents. If you have a word document and want to convert it to HTML, then a tool called word to HTML is available and it will automatically do the conversion. There are also web authoring tools which allow users to compose a web page without learning HTML in depth.

4.4.2 Web Pages with Tables

Designing web pages with tables is useful to display list of items available in a store, price of items and facilities to order them. We illustrate how a table is described in HTML with a simple example. In Figure 4.12, we show a web page with a table. The corresponding HTML description is shown in Figure 4.13. The table is specified between tags <TABLE BORDER = 1 CELLPADDING = 8> and </TABLE>. The table will be displayed with grid lines as BORDER = 1. The headers and cells will have some extra space as we have padded with CELLPADDING = 8. The first row specified by the tag <TR> has the column headings (Name, Pack, Price/pack – Rs., and shopping cart). Next row has the data for almond delight and the last row for cheese rich. When the anchor tag <AHREF= www.selectbakery.com/shopping cart.exe?biscuit=Almond delight>Add to cart </A> is clicked the web server hosting "www.selectbakery.com" will run a program called "shopping cart.exe" that will add the selected item to the shopping cart. (In the displayed page when *Add to cart* is clicked this is what will essentially happen in the underlying HTML description).

We conclude this section by summarizing HTML tags we have used (Table 4.1).

SELECT BAKERY-BISCUIT SHOP

Premium biscuits fresh from our oven

Name	Pack	Price/Pack Rs.	Shopping cart
Almond Delight	0.5	Rs. 80	Add to cart
Cheese Rich	1.0	Rs. 120	Add to cart

Figure 4.12 Web page with HTML table.

```
<HTML>
<HEAD>
<TITLE> Select Bakery - Fine baked products</TITLE>
<BODY BG COLOUR = "white" TEXT = "black">
<H1> Select Bakery Biscuit Shop </H1>
<P> Premium biscuits fresh from our oven </P>
<TABLE BORDER = 1 CELLPADDING = 8>
<TR>
<TH> Name </TH>
<TH> Pack </TH>
<TH> Price/Pack - Rs. </TH>
<TH> Shopping Cart </TH>
</TR>
<TR>
<TD> Almond Delight </TD>
<TD> 0.5 </TD>
<TD> Rs.80 </TD>
<TD> <AHREF = "www.selectbakery.com/shopping cart.exe?biscuit = Almond delight"> Add to cart </A></TD>
</TR>
<TR>
<TD> Cheese rich </TD>
<TD> 1.0 </TD>
<TD> Rs.120 </TD>
<TD><AHREF="www.selectbakery.com/shipping cart.exe?biscuit=cheeserich">Add to cart </A></TD>
</TR>
</TABLE>
</BODY>
</HTML>
```

Figure 4.13 HTML tags for displaying table.

Table 4.1 Basic HTML tags

Used for	HTML tags
Document	<HTML>htmldocument</HTML> <HEAD> doc.head</HEAD> <BODY> doc.body</BODY> <TITLE> doc.title</TITLE>
Text formatting	<B> bold </B> <EM> emphasis </EM> <H#> header </H > where # is 1 to 6 <FONT> text font

(Contd.)

Table 4.1 Basic HTML tags *(Contd.)*

Used for	HTML tags
Position, paragraphing	<CENTRE> centered content </CENTRE>
 line break <P> paragraph </P>
Table	<TABLE> Table content </TABLE> <TR> Table row </TR> <TH> Table header </TH> <TD> Table data </TD>
List	<UL> Unordered list </UL> <OL> Ordered list </OL> <LI> List item</LI>
Graphics	<IMG> graphic image <HR> horizontal rule/line
Linking	<A HREF> anchor content </A>

4.5 OTHER NETWORK SERVICES

We primarily emphasized the http protocol provided by a browser program interface and how it is used to search for documents and retrieve them from the web. In Figure 4.14, we give a conceptual organization of a browser which provides a uniform user interface to use other services available on the Internet.

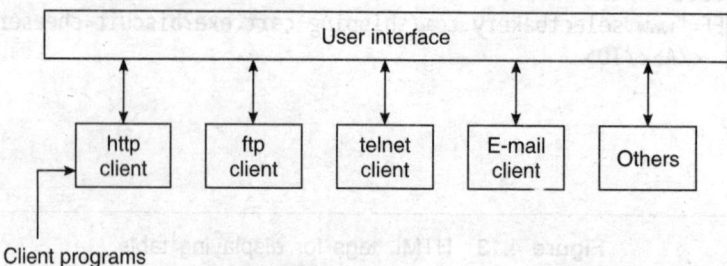

Figure 4.14 Various client programs allowed when you invoke a browser program.

The most important services are:

- *FTP*—a program to transfer files from any computer connected to the Internet, provided free access to the file is allowed.
- *Telnet*—a program that allows you to login to a remote computer (may be a compute server or database server) from your desktop or mobile computer provided, of course, you have access privileges. Telnet is very useful when you travel. You can log on to your computer from anywhere in the world using the Internet, read your mail, download any files you may need, and also run programs remotely.
- *E-mail*—sending/receiving mail on your client.

We will now describe how these programs are used.

4.5.1 File Transfer

When you transfer a file from a server to your computer, it is called *downloading* a file. On the other hand, if you send a file from your computer to another computer connected to the network, you are said to *upload* a file. FTP is an application that allows you to access any computer connected to the Internet and download file(s) from it. If you want to download a file named admission-rules from the web site of *x* college, you enter in your browser the information given in Figure 4.15.

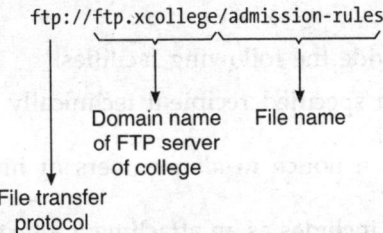

Figure 4.15 Use of FTP to download a file.

The remote file server (in this case *ftp.xcollege*) will send back a message asking for login id and password. If it is a publicly accessible file server, you normally write **anonymous** as user id and your e-mail id as password. Some systems will accept **guest** as password. It will also ask a filename to which the downloaded file should be transferred. Normally, it is better to receive a binary file image rather than a text file (in ASCII) unless you are certain that the text file can be opened as is on your computer. In windows environment the usual file extensions are *.zip* (for compressed binary file) or *.txt* (for ASCII text file).

4.5.2 Telnet

As we saw, by telnet we mean logging on to a computer from a remote computer and using facilities provided by it. If you type:

telnet: <domain name of the remote computer> or IP address (if you know it)

an application program in your browser will connect you to the remote computer and get ready to accept messages from it. The remote machine will send a message asking for your login id and password, exactly as though you are using that computer. You now give your login id and your password and it will admit you as an authorized user. You can have all the facilities such as running a program, accessing your files stored there, and modifying them if you so desire. You can even develop programs from a remote location, compile and execute them. If the Internet connection is via a high bandwidth line, you will not even know that you are remotely located.

Telnet is routinely used by software companies in India for the so-called "offshore software development". In this model, these companies often use the network of computers of the client remotely from India, upload programs, test and run them with live data. Errors are remotely corrected and any fault reports are attended to from India. This is highly economical as compared to sending software engineers to foreign countries to work on client sites. In fact, many software companies maintain existing systems for clients abroad from India. Sometimes a remote computer may be accessible to a group of people.

Access is controlled by a password and if it is a secure system more than one passwords. Further, when a team is working on a computer and need to access a database, it may be controlled depending on the user privileges. A group leader may be allowed to read data and write on the database, whereas some members may have read only privilege. Remote telnet is also used in some call centre operations. When a client complains about a problem in a software product, a call centre technical person can remotely log in, find out his or her problem, and solve it.

4.5.3 E-Mail

Current e-mail systems provide the following facilities:

- Send a message to a specified recipient technically called unicast or a group of recipients (multicast).
- A sender may send a notice to *all* members of his department. This is called broadcast.
- Send a message that includes as an attachment a text, audio, images or video file.
- Send a message along with a program which can be executed at the recipient's computer and send a response.

In order to be able to send or receive e-mail, you need a unique address. It is similar to the need to have a postal address if you want to receive letters. The e-mail address has the form:

yourname@name of your e-mail sever assigned by domain name server.

For example, my email address is:

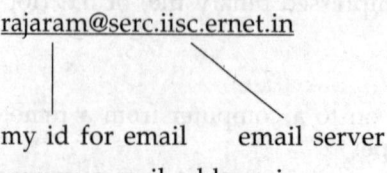

my id for email   email server

Another common e-mail address is

r-raman@gmail.com

Name of person   web e-mail provider's server name

In order to send/receive e-mail, you need an application program which should be installed in your computer. It allows you to compose a message, edit a message, specify the recipient or recipients by giving their e-mail addresses and specify names of attachment files (if any). It also provides a mail inbox for you to receive mail and an outbox to store mail to be dispatched from your machines. A typical e-mail is as shown in Figure 4.16.

If you are connected to a LAN in your company office, then there will be a mail server in the LAN and your mail boxes will normally be in that server. There is no need to dial up ISP. The organization's LAN will be connected to the ISP through a router and possibly a leased line of high bandwidth. The mail server will be on 24 hours a day, 7 days a week, and will be sending the mail almost immediately to the destination address (see Figure 4.17).

> Subject: Thank you
> From: ramu6@ee.iitk.ac.in
> Date: Wed, March 26, 2008 12:44
> To: rajaram@serc.iisc.ernet.in
> Cc: krishnan@cs.concordia.ca
>
> Dear Sir,
>
> Thank you for responding to my mail and answering my question.
>
> With regards,
>
> Sincerely, Ramu

Figure 4.16 A typical e-mail format.

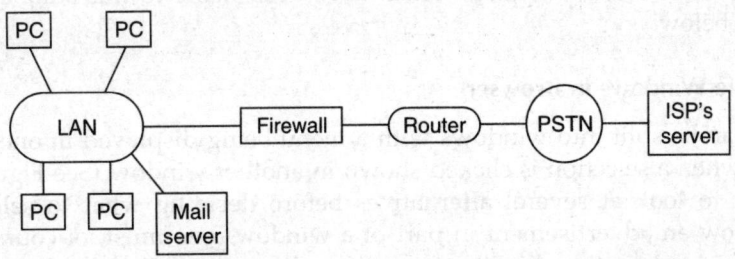

Figure 4.17 E-mail via a LAN.

If you are logged on and the server receives a mail, it will immediately forward it to you and a message will normally flash on your PC's VDU. The e-mail system on the Internet is normally very reliable. When an e-mail is sent from the outbox of the server, it is not erased immediately. The server waits for an acknowledgement from the intended receiver. If there is no acknowledgement for a specified period, the mail is resent. If there is no response again, the ISP informs the sender that the mail could not be delivered (it is normally said that the mail has bounced). A mail can bounce if a recipients e-mail id is wrong.

Mail systems also allow you to keep a directory of nicknames (or shorter names) for frequently used e-mail addresses. This reduces your typing effort. There are a number of other facilities available with e-mail system. Some of these are as follows:

- You can send the same mail to many by specifying all their email addresses in the *To* field. For example, you can write
  To: ramu@vsnl.com,kichu@sify.com,balu@sancharnet.in,seenu@iitm.ac.in and the same mail will go to ramu, kichu, balu and seenu.
- The same message can be broadcast by the ISP providing the e-mail service to all the subscribers.
- A message can be multicast, that is, broadcast to a restricted pre-specified group.
- E-mail can be forwarded to another e-mail address if it is specified.
- You can type the reply to the sender and send it by clicking on "reply". It is not necessary to retype the sender's address.

To summarize, e-mail has become a very important service provided by the Internet infrastructure and has proved fast, reliable and flexible.

Nowadays, many organizations and individuals use free e-mail service provided by companies such as Yahoo, Google, Microsoft, etc. A mail account is easy to open with any of these companies. The mail server and file space for each user is provided by these companies. For example, if you have an account with Google the e-mail id will be: your unique id@gmail.com. In order to access the e-mail account you access the web site of Gmail (usually bookmarked by you). It displays a window for you to enter your unique id and password. It then loads all your mail from the mail box and also the software to enter mail, edit it, send it, keep a file of sent mail and whole lot of other facilities. Normally such e-mail sites provide over 2 GB of mail box space.

4.6 ADVANCED WEB TECHNOLOGIES

So far we have described simple web pages where the whole page is accessed by the browser. There are several advances which have been made to this simple model. They are described below.

4.6.1 Multiple Windows in Browser

A web page can be split into windows with a menu being displayed in one window and the response when a selection is clicked shown in another window (See Figure 4.18). This allows a user to look at several alternatives before deciding what to select. It is also possible to show an advertisement in part of a window (one must, of course, make sure that the page is not cluttered with too much information). Technically this facility is known as *form*.

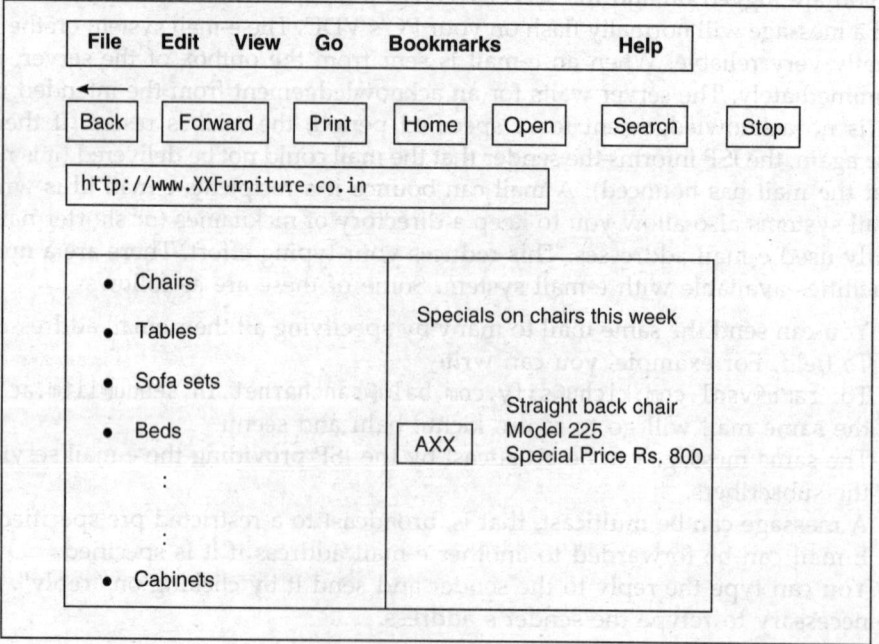

Figure 4.18 Windows in a browser display (Chairs clicked).

4.6.2 Common Gateway Interface Program

The web pages we have described are static. In other words, the web server stores a number of pages and the requested page is retrieved and displayed. During successive invocations, the displayed page does not change. There are many situations in which it will be advantageous to display varying information.

For example, when a user selects a URL a program in it can add 1 to a counter which increments whenever the web page is accessed. It can then send back to the browser a message such as:

"This site has been accessed 2556 times during the last month"

Such a message tells the popularity of the site. Another example is one in which available stock of various items are updated every time a purchase is made or new stock is added. It can then give the browser accessing it up to date stock position. Other common applications are up to the minute quotes of stock market prices, current weather conditions, etc. Such a program running in a server is known as *Common Gateway Interface* (CGI).

Another very useful application is to combine multiple windows in a browser with a program running in a web server. For example, when a customer selects an item which he or she wants to buy, a form can appear on a window requesting the buyer to enter his or her credit card number and the shipping address. This is input to the server. A program in the server uses this to create the shipping label and forward the credit card number to the credit card company for approval and payment.

There are several other advantages of being able to execute a program on a web server when it is invoked by a browser. These are:

- Personalizing advertisements knowing the items being purchased by a customer.
- Assigning a *shopping cart* to a customer when he or she starts selecting items for purchase. A shopping cart in e-commerce is a unique label assigned to a customer by the web server. This label is appended to each purchase made by a customer. If a customer collects items but does not purchase, the label can be retained along with the list so that if he or she visits the site again and adds more items the same unique shopping cart is available. Normally, the label (or in popular parlance virtual shopping cart) is preserved for a week or two after which it is erased.

4.6.3 Cookies

Another use of program running on a web server is popularly known as *cookie*. (I do not know why such a funny name is used). A cookie is a unique label assigned to a browser by the web server when it is accessed. This unique label is returned to the browser and stored there. When the browser revisits the URL, this unique label is sent to the server by the browser. The server uses this label and stores all the activity carried out by the browser during a visit. Thus, all purchases made, all advertisements shown, etc., to this user are remembered by the server. Thus, when the user revisits the site later, the web server can personalize the information sent to the browser. Thus, each user gets a different piece of information depending on past pattern of their use of the web server. Cookie technology intrudes on the privacy of users by remembering their activity. Thus, most browsers give a choice to the user to either accept or reject a cookie. If you do not want the server to remember what you did earlier, you should reject the cookie. This will preserve your privacy. On the other hand, if you frequently visit an e-bookshop, for example, it may be advantageous for the book store to give you personalized information on new books received which may interest you, book reviews, etc. The final choice should, of course, be yours.

4.6.4 Active Documents

The browsers as we have seen so far are passive. They are primarily used to display web pages. Browsers are run on client computers. Thus, they can also execute programs. If a web server sends a program to run on the browser along with a web page, it is called an *active document*. The only problem is that the client machines are not identical. They are made by different manufacturers and may run different operating systems. Thus, the program should be truly machine independent. Thus, active documents are Java programs. The Java programming language is very rich with a large number of features and requires good professional programming skills. Thus, a simple subset of Java called *Java Script* is used for active documents. Java Script can be embedded in a standard HTML file. When the browser encounters Java Script in the HTML document, it executes the program and displays the results on the browser screen.

Active documents have several uses. For example, if an e-shop wants to demonstrate how a gadget is to be used, then an animation program of the demonstration can be sent to the browser for execution. If the video clip plays in the server, the browser will have difficulty receiving it as the Internet is not designed for receiving video. Further, if several hundred browsers ask for the same video clip, the server may not be fast enough to satisfy the requests.

4.6.5 Cascading Style Sheets

Many HTML tags have associated properties which are set by default by browsers. For example, <H2> tag has a font size property for which browsers use a default value of 30 points. To allow freedom to a designer of a web page there should be a method of specifying the font size. This is the purpose of style sheet. The reason it is called cascading style sheets is because it is a sequence of standards CSS1, CSS2, CSS3 defined by the World Wide Web consortium (W3C). These standards are implemented by browser designers. Each standard adds new features.

There are three levels of style sheets called *inline, document* and *external*. Inline style sheets apply to the contents of a single tag. Document level style sheets apply to the whole body of a given document. External style sheets can apply to the bodies of any number of documents. Inline style sheets have precedence over document style sheets which have precedence over external style sheets.

A simple example of style in HTML document is shown in Figure 4.19.

This will print heading 1 in Times Roman, 24 point in bold face and centre it. The next line will print heading 2 in courier 18 points bold face. This is an inline style. To make it document level, we can define it at the top to be used in the document. If we put all the styles in a file and invoke it, then it can be used for a set of documents. For details you may read textbooks on HTML.

```
<!.....fonts.html... An example to illustrate . . . >
<html>
<body>
<centre>
<h1 style = "font-family'Times Roman';
         font size:24pt;font-weight:bold"></centre>>
Essentials of E-commerce
</h1>
<h2 style ="font-family:'courier';font-size:18pt;
         font-style:italic;font-weight:bold">
Introduction
</h2>
</body>
</html>
```

Figure 4.19 A style sheet.

4.7 CONCLUSIONS

In this chapter, we primarily discussed the technological aspects of network services. The discussion was centred around the World Wide Web and its use in e-commerce. Another important aspect of the World Wide Web is concerned with how to launch an e-business, and design appropriate web sites. This is more of a management issue which includes questions such as aesthetics, visibility and services to be provided. We will discuss these aspects later in this book.

SUMMARY

1. Hypertext is a method of organizing a collection of textual data in such a way that they are linked by specified keywords.
2. In hypertext-based search, if a keyword is selected, it allows you to retrieve all texts which are linked by specified keywords.
3. World Wide Web (www) is an application program which uses the Internet infrastructure to organize and link information services.
4. The World Wide Web consists of:
   (i) Documents called web pages in which keywords are marked and linked to other web pages. A language called HyperText Markup Language (HTML) is used to describe web pages.
   (ii) A protocol called HyperText Transfer Protocol (HTTP) for accessing hypertext data from remote computers.
   (iii) A unique address for each web page called URL. URL consists of www followed by the domain name of the server hosting the web page and a path name to requested item in the web page.
5. A program called a browser allows one to view the contents of specified URLs. Browsers have very good graphical user interface to simplify retrieval of information available on the World Wide Web.

6. Tools called search engines are available to search for specified information on the World Wide Web. An expression consisting of keywords of interest connected by Boolean connectives such as AND, OR, and NOT are used to specify to the search engine information requirement.

7. Search engines index all the web pages using keywords, and these are used to retrieve information from the web. Indexing is done by software known as web crawlers which periodically visit all web pages to collect keywords and index them.

8. Browsers also provide other useful tools. Some of these are to:
    (i) FTP to access a file from a remote computer and download it on a computer
    (ii) Telnet to log on to a remote computer and use it
    (iii) Send and receive e-mail

9. Markup languages add tags to documents which assign special meanings to various character strings used in the document. In other words, markup languages describe the structure and content of documents.

10. A markup language called Standard Generalized Markup Language (SGML) was designed as an international standard but is not widely used due to its complexity. However, two markup languages derived from SGML, called HTML and XML, are widely used.

11. HyperText Markup Language (HTML) is used to create visually pleasing display of documents on the VDU of a computer connected to the Internet.

12. The primary application of HTML is to add markups to documents stored in a World Wide Web page. In addition to formatting the document as paragraphs, displaying specified words as bold face or italics and generally improving the appearance of the document when displayed, these markups link the document to one or more other documents, pictures, etc. Tables can also be created using HTML and displayed.

13. There are several other facilities provided by browsers, the most important being FTP, telnet and e-mail.

14. For transferring programs or other files available on the World Wide Web, a program called FTP (File Transfer Protocol) is used which is provided by browsers.

15. Browsers also allow a user to log on and execute programs in any other computer connected to the web provided the user has the requisite password. This is called telnet.

16. Electronic mail systems are supported by browsers.

17. A web page may be created with multiple windows. One window may be used to show the menu and the other to show the response when one of the menu items is selected.

18. It is possible for a web server to execute programs. These programs may be invoked when a browser selects the URL of the web page. Such a program invocation allows time varying information (such as stock prices, current inventory levels, etc.) to be displayed on the browser. Such a program is called Common Gateway Interface (CGI).

19. A web server can assign shopping cart (which is a unique label attached to a customer's purchase) to each customer. This shopping cart can be used by the same customer during several visits within a specified period.
20. A cookie is a unique label attached to a customer which is used to store in the browser a history of a customer's past purchases, searches, etc. This personalizes the browser's behaviour for each customer. A cookie may be rejected by a customer if he or she values his or her privacy.
21. A web server can send a program to run on a browser invoking it. Such a program is normally written in a language known as Java Script. This is known as active document. Among several uses of active document, one is to obtain from a server video clips such as how to use a gadget sold to a customer.
22. Appearance of a web page can be tailor-made by using style sheets.

EXERCISES

4.1 Define information browsing using the Internet. What facilities does a browser provide?
4.2 What is hypertext? What is hypertext-based browsing? Give an example of hypertext-based browsing.
4.3 What is the main advantage of hypertext?
4.4 Who first came up with the idea of World Wide Web? Where was this implemented first?
4.5 What do you understand by World Wide Web? What are the different steps needed to create the World Wide Web?
4.6 What is URL?
4.7 What are the different parts of a URL? Explain the purpose of each part.
4.8 What is a web browser? Name two popular web browsers.
4.9 What is a web page? How is a web page accessed?
4.10 What is a bookmark in a browser? Why is it useful?
4.11 What is a search engine? What are the main principles used by a search engine to retrieve information from the web? Give the names of some popular engines.
4.12 Enumerate some useful tips to search for information desired by you on the World Wide Web.
4.13 What are the various client programs which can normally be invoked from a browser program?
4.14 What is FTP? How do you use FTP?
4.15 How do you download a file using FTP?
4.16 What is telnet? How is it used?
4.17 What facilities does an e-mail system provide?
4.18 What are the different parts of an e-mail address?
4.19 When you send an e-mail, do you receive an acknowledgement? How do you know that the e-mail has been delivered to the intended recipient?
4.20 What do you understand by the term *markup*? What is SGML? What is the structure of SGML?

4.21 What is HTML? What are its applications?
4.22 Write using HTML the following document:
GONE WITH THE WIND
Margaret Mitchell
RANDOM HOUSE
New York, USA
4.23 Can image files be embedded in an HTML document? If yes, how?
4.24 What is a hyperlink? How do you link an image to an HTML document?
4.25 Can multiple windows be created on a web page? If yes, what are the applications?
4.26 What do you understand by the term forms in relation to web pages?
4.27 What is a Common Gateway Interface (CGI)? What are its uses?
4.28 What is a shopping cart in e-commerce? What is its main use?
4.29 Can a shopping cart be used during several visits to a web server? Explain how it is implemented by a web server.
4.30 What is a cookie? What are its uses?
4.31 Discuss the pros and cons of cookies.
4.32 What do you understand by an active document delivered by a web server? What are some of its uses?
4.33 What is Java Script? Where is it used?
4.34 What is cascading style sheet? What is its use?

OBJECTIVE QUESTIONS

Each question has four possible answers. Pick the most appropriate answer.

4.1 A World Wide Web is
 (a) Another name for the Internet
 (b) A worldwide connection of computers
 (c) A collection of linked information residing on computers connected to the Internet
 (d) A collection of worldwide distributed information.
4.2 A World Wide Web contains web pages
 (a) Residing in many computers
 (b) Created using HTML
 (c) Linked to other web pages
 (d) Which are related and are linked together using specified selectable keywords
4.3 A web page is located using
 (a) Universal Resource Links (b) Universal Resource Locator
 (c) Universal Record Links (d) Uniform Resource Links
4.4 A URL is
 (a) Used to download a file
 (b) Used by a browser to locate a web server
 (c) A bookmark to create a web site
 (d) The IP address of a web server

4.5 The protocol used by the World Wide Web is called
   (a) Hyperlink to transfer web pages
   (b) Hypertext marking protocol
   (c) Hypertext transfer protocol
   (d) Hypertext transmission protocol

4.6 By hypertext, we mean
   (a) A set of related documents
   (b) A set of document linked by selectable keywords
   (c) A text which promotes non-linear reading
   (d) An enormously large text with embedded multimedia information

4.7 A web page is a
   (a) Page stored on the World Wide Web
   (b) Page stored in a web server
   (c) Page of a large text stored in a web server
   (d) Formatted annotated document stored in a computer with links to several other documents

4.8 A URL specifies the following:
   (i) Protocol used
   (ii) Domain name of server hosting web page
   (iii) Name of folder with required information
   (iv) Name of document formatted using HTML
   (v) The name of ISP
      (a) (i), (ii), (iii), (iv)   (b) (ii), (iii), (iv), (v)
      (c) (i), (iii), (iv)   (d) (i), (ii), (iii), (v)

4.9 A browser is a program used to obtain
   (a) Information
   (b) Web pages
   (c) Web pages for specified index terms
   (d) Web pages with specified URL

4.10 HTML stands for
   (a) HyperText Making Links
   (b) HyperText Markup Language
   (c) Higher textual Marking of Links
   (d) HyperText Mixer of Links

4.11 HTML is similar to a
   (a) Word processing language
   (b) Screen editor
   (c) Scripting language
   (d) Search engine

4.12 A search engine is
   (a) A program used to search for documents
   (b) An engineering artifact to search for specified files
   (c) A program to search for desired information from the World Wide Web
   (d) Used in e-commerce

4.13 A search engine
   (a) Uses a Boolean query to search the World Wide Web
   (b) Searches web using search terms
   (c) Uses keyword provided by a user on a browser to select appropriate documents from the World Wide Web
   (d) Uses URLs to search web pages

4.14 Desirable properties of a web site are
   (i) A meaningful address
   (ii) Help and search facilities
   (iii) Links to related sites
   (iv) Features to allow users to give feedback
   (v) Hosting on a mainframe
      (a) (i), (ii), (iii)
      (b) (i), (ii), (iii), (iv)
      (c) (i), (ii), (iii), (iv), (v)
      (d) (i), (ii), (iii), (v)

4.15 A web crawler is a
   (a) Program to crawl through several web pages
   (b) Language to develop search engines
   (c) Program which systematically visits web pages linked by keywords and creates an index useful for search engines
   (d) System to search the World Wide Web with a set of keywords

4.16 For successfully searching the World Wide Web, it is preferable to use
   (i) URL of web servers
   (ii) Use narrow search terms to limit retrieved documents
   (iii) As many search terms as possible
   (iv) E-mail address of a web page
      (a) (i), (iii)
      (b) (i), (ii)
      (c) (ii), (iii)
      (d) (ii), (iii), (iv)

4.17 Bookmarks are
   (i) URLs of useful web sites which are used often by a user
   (ii) The results of web search
   (iii) Used by regular users of the World Wide Web
   (iv) An aid to remember context during search
      (a) (i), (ii)
      (b) (i), (iii)
      (c) (i), (iv)
      (d) (i), (iii), (iv)

4.18 The expansion of the acronym SGML is
   (a) Standard General Machine Language
   (b) Standard Generalized Markup Language
   (c) Super General Machine Language
   (d) Simple Good Markup Language

4.19 The primary purpose of SGML is to
   (a) Formally describe structures and contents of documents with appropriate notation
   (b) To make a document machine readable
   (c) Formally describe web pages
   (d) Describe formally documents

4.20 SGML laid the foundation for
   (a) HTTP
   (b) HTML and XML
   (c) XML
   (d) Several application specific markup languages

4.21 An anchor tag is used in HTML to
   (a) Define a hypertext
   (b) Add an image to a web page
   (c) Add a table to a web page
   (d) Create links to other web pages on the World Wide Web

4.22 For including an image in a web page, we use
   (a) An anchor tag with an image in it
   (b) A pair of tags <IMG> and </IMG> enclosing the image
   (c) A file tag IMG followed by the source of the image file and filename
   (d) A tag named <IMAGE> with end tag </IMAGE>

4.23 The tag used to include an unordered list in a web page is
   (a) <UOL> and </UOL> at the end of the list
   (b) <UL> and </UL> at the end of the list
   (c) <L> and </L> at the end of the list
   (d) <LIST> and </LIST> at the end of the list

4.24 Tables can be introduced in web pages using a
   (a) Set of tags <TABLE>, <TR>, <TH> and <TD>
   (b) Set of tags for rows and columns
   (c) Set of tags <TABLE>, <ROW>, <COL>, <VAL>
   (d) Set of user-defined tags

4.25 Cascading style sheets in HTML are used to define
   (a) A cascade of formats from which any can be chosen
   (b) A set of formats to represent the same web page
   (c) A user-defined style sheet template for tailor-made appearance of a web page
   (d) Separate formatting methods for a web page

4.26 Browsers provide besides HTTP the following:
   (i) FTP
   (ii) Telnet
   (iii) E-mail
   (iv) DTD
   (a) (i), (ii)
   (b) (i), (iii)
   (c) (i), (ii), (iii)
   (d) (i), (ii), (iii), (iv)

4.27 FTP is used to
   (a) Download files from the web server to the browser
   (b) Transfer files from the browser to the web server
   (c) Transfer files between web servers
   (d) Specify files stored in web server

4.28 Telnet allows you to
   (a) Enable peer to peer computing
   (b) Use any computer in the world from your PC

(c) Log on to a remote computer connected to the Internet from your computer
(d) Log on and use a remote computer whose IP address/domain name is known to you if permitted

4.29 CGI is
(a) A program which is executed in a web server
(b) A program which is executed by a browser
(c) Common Gateway Interface to any computer
(d) Common Gateway Interaction between two browsers

4.30 CGI is primarily used to
(a) Personalize a web server
(b) Provide time varying information to browsers
(c) Log every use of a browser
(d) Make a server machine independent

4.31 Some of the uses of CGI are
(i) Assigning a shopping cart to a customer
(ii) Personalizing advertisements appropriate to a customer
(iii) Allowing a user to provide information such as a credit card number and address to web server
(iv) Sending a video clip to a browser
    (a) (i), (ii)   (b) (i), (ii), (iii)
    (c) (i), (ii), (iii), (iv)   (d) (i), (iii), (iv)

4.32 A cookie is a
(a) Program which runs on a browser to debug its program
(b) Unique label assigned to a browser by a web server when it is accessed
(c) Spy program to detect frauds
(d) Spy program to monitor a user's activities

4.33 A cookie is used to
(a) Log a browser's usage pattern when it accesses a web server
(b) Monitor illegal uses by a browser
(c) Personalize a browser
(d) Detect viruses in web server programs

4.34 An active document is
(a) A document which is time varying
(b) A web page sent to a browser by a server
(c) A program sent to a browser by a web server which can be executed in the browser
(d) An active web page sent to a browser by a web server

4.35 Java script is a
(a) Subset of Java which is machine independent
(b) Subset of Java used to create active documents
(c) General purpose object oriented programming language
(d) Scripting language based on Java

4.36 Active documents may be used to
(a) Send live animated demos to browser by a web server
(b) Send web server activities to a browser
(c) Send time varying information to a browser
(d) Spy on the activities of a browser

CHAPTER 5

Secure Messaging

LEARNING GOALS

In this chapter, we will learn:
1. About security threats when the Internet is used.
2. Encryption methods to ensure security of messages sent using the Internet.
3. About digital signatures and their importance.
4. Ensuring overall security of e-commerce site.

5.1 INTRODUCTION

Security of transactions between customers and vendors in e-commerce is of paramount importance. There are several types of threats which may arise. Some of these are:
1. Attempt to access a web site and modify/destroy its contents.
2. Attempt to access a web site and read confidential information such as credit card numbers and company confidential payroll data.
3. Send malicious programs such as viruses, worms and Trojans to a web server by a browser.

We have already described in Chapter 3 security measures taken using various types of firewalls to counter these threats. These are not sufficient in e-commerce. There are other requirements such as confidentiality of message sent by a customer to a vendor, authentication of message, non-repudiation of messages sent and integrity of message. We will now see the types of problems one faces when e-documents are transmitted via an unprotected messaging medium such as the Internet. The scenario is depicted in Figure 5.1.

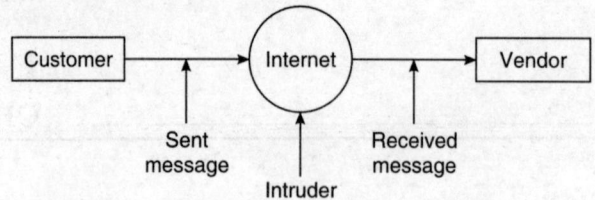

Figure 5.1 Messaging scenario in e-commerce.

The following problems could arise in sending messages:

1. Customer may send a purchase order to a vendor along with confidential information such as a credit card number. Intruders may be able to read this information and copy it. Thus, we have to ensure that intruders are not able to read the message and even if they are able to read the message they are not able to decipher it and get useful information such as a credit card number. In other words, the *confidentiality* of messages should be preserved.
2. A customer may send a purchase order to a vendor. Intruders may intercept it and alter it and forward it or may even replace it by altering the shipping address to their address. A vendor should be able to make sure that the order was indeed sent by the customer and that it has not been altered. In other words, the *authenticity* of message sent by a customer has to be ensured.
3. A customer may send a purchase order to a vendor and later claim it was not sent by him or her, or that the items ordered and the quantity supplied by the vendor are not what he or she ordered. In other words, a vendor must be sure that a customer does not *repudiate* the actual order.
4. A customer may order some shares from a stock broker at time T1. When a bill is received he or she may claim that he or she ordered it at time T2 at which time the share prices may have gone down by 20% and agrees to pay only the smaller amount. This example shows the importance of *time stamping* of a transaction and authenticating it.

All the above problems require methods of preventing such events and if unsuccessful recover from them. The primary method of achieving this is by what is known as *encryption*. All messages exchanged between participants in e-commerce are encrypted. In this chapter, we will describe various methods of encryption and their use in ensuring secure exchange of messages in e-commerce.

5.2 SYMMETRIC DATA ENCRYPTION WITH PRIVATE KEY

We start by defining some terms we will be using. An original message is known as *plain text*. A coded plain text is known as *ciphertext*. The process of coding plain text to ciphertext is known as *encryption*. The various methods of encryption constitute the subject area of *cryptography*. Methods used for decoding ciphertext without a full knowledge of the key(s) used for encryption is commonly known as *cryptanalysis*. The primary aim of cryptography is to be able to withstand cryptanalysis attacks on the ciphertext and preserve the secrecy of plain text.

In Figure 5.2, we show the general scheme used in private key encryption. Referring to Figure 5.2 *encryption algorithm* is a mathematical transformation of plain text to ciphertext using a *secret key* k. The encryption algorithm uses the key to transform plain text in such a way that the ciphertext cannot be understood by a evesdropper who reads it. Different keys will produce different ciphertexts for a given plain text. The ciphertext is decoded using a *decryption algorithm* which also uses the *same* secret key k. This is called *symmetric key cryptography*.

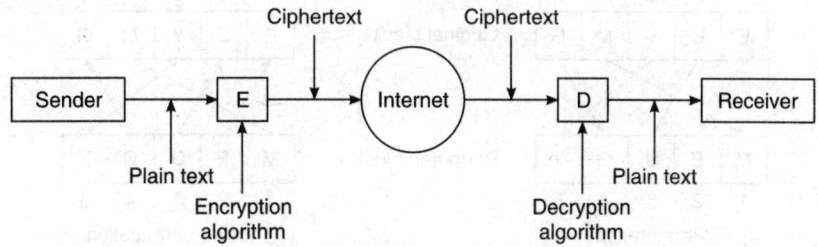

Figure 5.2 Block diagram of symmetric private key encryption.

The primary requirements of symmetric key encryption are:
1. The encryption algorithm is not secret. It is known to all. The encryption key is secret. The same key is used by both the sender and the receiver.
2. The sender should be able to communicate to the receiver the secret key by using a secure channel other than that used for sending the message.
3. A cryptanalyst who is able to obtain several pieces of ciphertext should not be able to guess the key. (As the encryption algorithm is normally public knowledge).
4. Plain text in e-commerce is a string of bits which may represent alphanumeric character strings, graphs, audio or video.

A popular encryption algorithm encrypts blocks of plain text using two transformations commonly known as confusion and diffusion. A *substitution* of blocks of bits by another block is used as confusion. A *permutation* of bits is used as diffusion. We will illustrate the method with an example using a string of English letters.

Suppose a plain text is:

AQUICKFOXJUMPST

The first transformation is substitution, that is, replacing a character by another character. We will assume the 26 letters are arranged circularly and we replace a character by the fourth character following it in the collating sequence (i.e., alphabetical order). In other words, A is replaced by E, B by F, ..., Y by C and Z by D. Applying this substitution the transformed plain text is:

Ciphertext 1:   EUYMGOJSBNYQTWX

This string is now divided into blocks of characters. The block size is a designers choice. We choose blocks of five characters. These five characters are transposed using a permutation of characters with a key. If (41253) is the permutation, it is interpreted as: [See Figure 5.3(a)] replace the first character of this five-character block by the fourth character, the second character by the first, the third character by the second character, the fourth by the fifth and the fifth by the third. The transformed ciphertext 2 is obtained as shown below:

Ciphertext 1 blocked: EUYMG | OJSBN | YQTWX |

Transposition by permutation gives:

Ciphertext 2: MEUGY | BOJNS | WYQXT

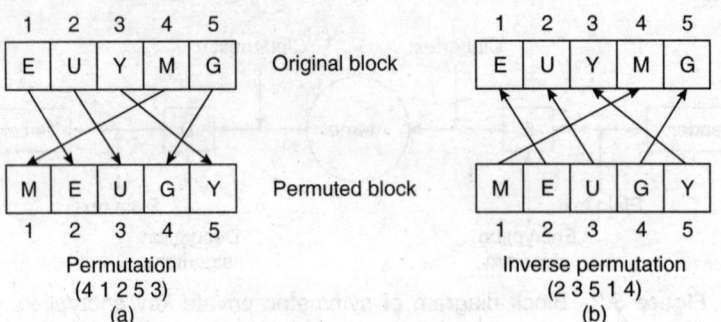

Figure 5.3 Permutation transformations used in encryption and decryption.

This is the final ciphertext.

Decryption of the ciphertext is done by applying the inverse transformation on it using the specified key. Observe from Figure 5.3(b) that inverse permutation is obtained by inverting the arrows in the original permutation. Inverse permutation is thus (23514).

Given a ciphertext, we will now see how to decrypt it. Given a ciphertext MEUGYBOJNSWYQXT.

We apply the inverse permutation (23514) to blocks of 5 characters. This gives:

EUYMGOJSBNYQTWX

Replacing a letter by the fourth letter preceding it we get:

AQUICKFOXJUMPST

which is the original plain text.

It is very difficult to guess the plain text from a given ciphertext unless one knows the decryption key. The algorithm on how the encryption is done can be publicized without compromising the encrypted message as a person has to know the transcription and transposition keys to be able to obtain the plain text given the ciphertext. If the transcription and transposition keys are changed frequently, then even if a number of ciphertexts are obtained, it will be difficult to decode them and get the plain text.

The basic idea of using substitution and permutation is used in a standard method for encrypting *binary strings*. This is called Digital Encryption Standard (DES) which we describe next.

5.2.1 Digital Encryption Standard

This encryption method was first proposed by IBM in 1975 and standardized in 1977. It was a standard method of encryption for a long time but with increase in speed of computers, it is no more considered secure as a cryptanalyst can break the code (i.e., find the encryption keys) by exhaustively searching for all the keys using a fast computer. However, a modification of DES called triple DES is now used which is more secure and is difficult to break.

DES is called a block cipher as the input bit string is divided into 64-bit blocks and each 64-bit block is transformed using the same key. DES is also called a symmetric encryption algorithm as the same key is used for encryption as well as decryption.

The algorithm used by DES is designed in such a way that the operations performed for encryption and decryption can be easily performed by hardware circuits. Thus, integrated circuit chips can be designed to perform DES encryption of binary strings fed to it.

DES transforms blocks of 64 bits corresponding to binary encoding of ASCII characters of message text. The algorithm uses exclusive OR operation defined by:

$A \oplus B = A \cdot \overline{B} + B \cdot \overline{A}$ (See Table 5.1) where $\oplus$ is the exclusive OR operator. Exclusive OR operation is used as it has the interesting property given by the equation:

Table 5.1 Exclusive OR Operation

A	B	A $\oplus$ B
0	0	0
0	1	1
1	0	1
1	1	0

If $C = P \oplus K$, then

$$P = C \oplus K$$

where P is used for plain text, K is the key and C is ciphertext.

Thus, if a plain text is:

P = 01101100 11011000 11011010

and key K = 10101111 00101100 01011011

Encrypted message is obtained by bit-wise exclusive ORing string P with string K. We thus get

$C = P \oplus K$ = 11000011 11110100 10000001

$C \oplus K$ = 01101100 11011000 11011010

which is the original plain text P.

The actual DES encryption algorithm is quite complex. We will describe the basic idea using Figure 5.4. DES encoding is applied to blocks of 64 bits of the given plain text. 64-bit blocks are encrypted using a 56-bit key. After an initial permutation of a 64-bit block, the block is divided into a left and a right block each 32 bits long. The left and right blocks are represented by L and R respectively. Sixteen rounds of identical operations are applied using a different key in each round on the right and left halves of the block. After sixteen rounds the right and left halves are joined and a final permutation which is the inverse of the initial permutation is applied to the 64-bit block giving the final encrypted output.

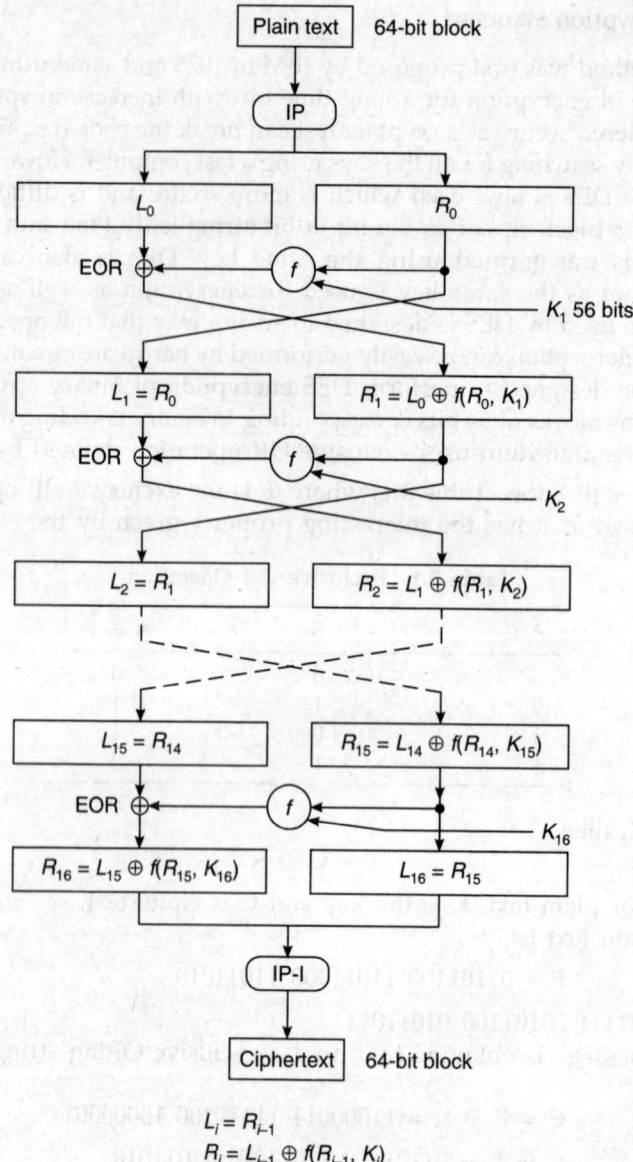

Figure 5.4 DES encryption of 64-bit block.

Figure 5.4 pictorially illustrates how the algorithm works. Referring to Figure 5.4 after initial permutation plain text is divided into left and right halves represented by L_0, R_0 respectively. L_i and R_i represent in the ith round value of L and R. K_i is the key used in the ith round. The operation shown in Figure 5.4 is very complex. The following steps are followed (See Figure 5.5.):

Step 1: Choose 48 bits from a 56-bit key (In each round a different set of 48 bits are chosen so that the key used in each round K_i is different). A well-defined key schedule is essential.

Step 2: Expand R_{i-1} to 48 bits by table lookup giving $E(R_{i-1})$.

Step 3: Exclusive OR $E(R_{i-1})$ with K_i.

Step 4: Compress the 48 bits result to 32 bits using substitution with table lookup.

Step 5: Exclusive OR result with L_{i-1} to get R_i.

The whole process is represented by the equation $R_i = L_{i-1} \oplus f(R_{i-1}, K_i)$ This algorithm is depicted in Figure 5.5 and is called *Fiestel cypher*.

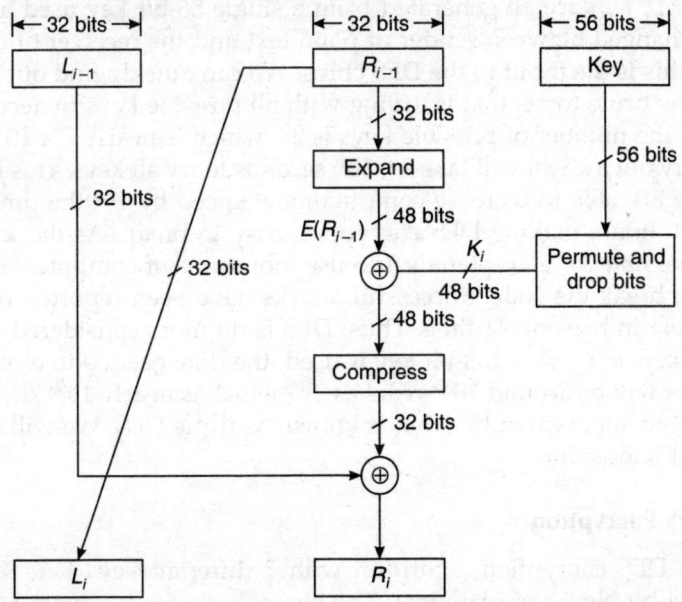

Figure 5.5 The operation performed by *f* of Figure 5.4 (one round).

The key transformation from 56 bits to 48 bits in each round, extension of bits and contraction of bits are all done by a set of tables whose details we do not discuss. (Interested readers can read the book by W. Stallings given in the set of references). At the end of *i*th round, we get

$$L_i = R_{i-1}$$
$$R_i = L_{i-1} \oplus f(R_{i-1}, K_i)$$

In the last round, we get $L_n = R_{n-1} \oplus f(R_{n-1}, K_n)$. R_n is appended to L_n giving the final encrypted 64-bit string.

The process of decryption of DES ciphertext is essentially the same as the encryption process. To decrypt, the ciphertext is taken as the input to the DES algorithm. The keys $K_1, K_2, ..., K_{16}$ are, however, used in the reverse order. That is, use K_{16} in the first round, K_{15} in the second round and K_1 in the last round. This is a very good feature of DES because it means that we need not implement two different algorithms to encrypt and decrypt. If an integrated circuit chip is designed to implement DES encryption, then the same chip can also be used to decrypt with the same primary key K from which $K_1, K_2, ..., K_{16}$ are derived. Figure 5.6 illustrates this.

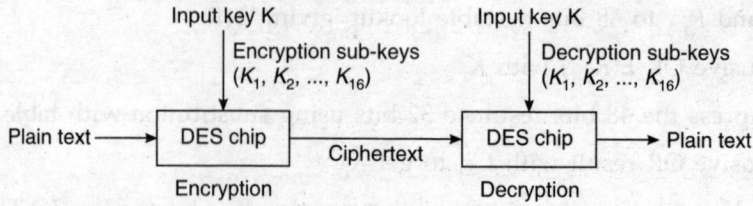

Figure 5.6 Encryption and decryption with DES chip.

As $(K_1, K_2, ..., K_{16})$ are all generated from a single 56-bit key used at the beginning, the only key exchanged between sender of plain text and the receiver of ciphertext is this master key and this is the input to the DES chips. We can quickly find out the time needed to find this key by brute force, that is, trying with all possible keys to decode a ciphertext. With 56-bit keys the number of possible keys is 2^{56} which is nearly 7×10^{16}. If it takes one microsecond to try out a key it will take 7×10^{10} seconds to try all keys. This is approximately 1100 years. If we are able to increase computational speed by 10^6 the time will reduce to approximately 10 hours making DES encryption easy to break. As the keys can be tried in parallel it is possible for a cryptanalyst to use thousands of computers connected to the Internet to try to break the code. Successful attacks have been reported of decoding DES encoded ciphertext in reasonable time. Thus, DES is no more considered very secure. We need larger size key. If $3 \times 56 = 168$-bit key is used, the time needed to break the code with the best computer will be around 10^{30} years as 2^{168} which is nearly 10^{50} keys are to be tried. Thus, DES has been superseded by what is known as triple DES. We will describe "Triple DES" in the next subsection.

5.2.2 Triple DES Encryption

Triple DES uses DES encryption algorithm with 3 different keys KA, KB and KC each 56 bits long on 64-bit blocks of plain text. The general scheme is shown in Figure 5.7. The use of the decryption block D in the middle has no cryptographic significance as the algorithm used by E and D are the same and using KA with the first block and KB with the second block will essentially encrypt the plain text with two independent keys.

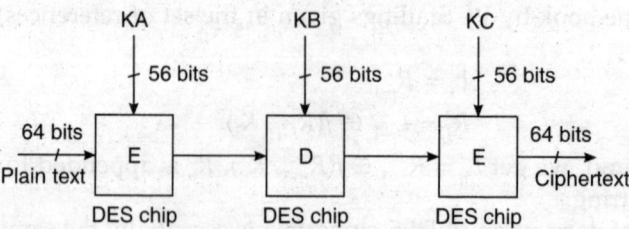

Figure 5.7 Triple DES encryption.

The main reason D is used as the intermediate block is to ensure compatibility with single DES. If KA = KB, then Figure 5.7 is equivalent to single DES with key KC. As we have pointed out in the last section using 3 independent 56-bit keys is equivalent to using a 168-bit key and makes this encryption practically unbreakable with the best computers of the foreseeable future. As it uses existing DES algorithm, it is easy to implement in hardware. Currently Triple DES abbreviated to 3DES is the most popular symmetric single key encryption method.

5.2.3 Advanced Encryption Standard

As DES was becoming insecure, the National Institute of Standards and Technology (NIST) of USA decided to introduce a new standard. In mid-1997 NIST called for worldwide participation to propose a new secure encryption system which will be accepted as a world standard. The important conditions specified by NIST were:

- It must be an open algorithm and publicly disclosed. Security must be ensured by the key only.
- It should be available to all without having to pay royalty.
- It should be a symmetric key method.
- It should be a block cipher with 128-bit blocks.
- It should be simple enough to be realized in hardware and implementable even in credit cards.
- It should not have a secret entry (called *trap door*) known to the designer which can be used to break the code.
- It should use keys of length 128, 192 or 256 bits.

By mid-1998 many proposals were received. Fifteen of these were chosen and publicized for cryptographers from all over the world to examine and try breaking it to find its strength. Based on their comments, finally in mid-2000, NIST announced the algorithm proposed by two Belgian cryptographers—Dr. Vincent Rijmen and Dr. Joan Daemon. This was called *Rijndaal* (pronounced Rain doll) *algorithm* which is now the latest Advanced Encryption Standard.

The encryption method is simple but quite sophisticated as it uses a schedule of keys based on deep mathematical theory. The full method is published in the web site of NIST with URL http://www.nist.gov/aes/ for those interested in examining it in detail and understanding the theory. We give as Algorithm 5.1 the gist of AES algorithm. This is primarily intended to illustrate the basic simplicity of the algorithm.

The algorithm is a block cipher which works with 128, 192 or 256 bit-blocks with 128-bit keys. A 128-bit block consists of 16 bytes. The algorithm for the case of 128-bit block and key is as follows:

Algorithm 5.1   AES Encryption

Step 1:   Divide the plain text into 16-byte blocks ($a_0, a_1, a_2, ..., a_{15}$)

Step 2:   Arrange each block as (4×4) byte matrix called state S

$$S = \begin{pmatrix} a_0 & a_4 & a_8 & a_{12} \\ a_1 & a_5 & a_9 & a_{13} \\ a_2 & a_6 & a_{10} & a_{14} \\ a_3 & a_7 & a_{11} & a_{15} \end{pmatrix} \begin{matrix} \text{Row 1} \\ \text{Row 2} \\ \text{Row 3} \\ \text{Row 4} \end{matrix}$$

Step 3: $S \leftarrow S \oplus$ key (0) where key (0) is a 4×4 byte key matrix. The exclusive OR is bit-wise exclusive OR *for* i = 1 to 9 *do* the following:

Step 4: S ← SUB(S) where SUB is a substitution operation (which is non-linear) in which each byte of S is substituted by a byte of a (16 × 16) substitution matrix by table lookup

Step 5: S ← left circular shift row (i) by (i − 1) bytes

Step 6: S ← Mix bytes in columns of S by different amounts. (This is a special operation which is a linear combination with predetermined weights)

Step 7: S ← S ⊕ key_i, where key_i is a 4 × 4 byte matrix and is a function of i

end for

Step 8: S ← SUB(S)

Step 9: S ← left circular shift row_i by i − 1 bytes

Step 10: S ← S ⊕ key_{10}

Step 11: CT ← S

Step 12: Rearrange matrix CT columnwise to get 128 block of ciphertext

Even though the decryption key schedule is the same for both encryption and decryption, the sequence of transformations needed to obtain the plain text from the ciphertext is different. This implies that the software used for decryption is different from that used for encryption. If we implement the encryption algorithm in hardware (IC chip) a different hardware chip is needed for decryption. AES software works reasonably fast on a variety of processors.

AES document gives several suggestions for efficient implementation of the standard on 8-bit processors used by smart cards and 32-bit processors used in PCs. In our discussion we have only considered essentials of AES. Those interested in more details may refer to the book by Stallings given in the set of references or refer to www.nist.gov/aes/.

5.3 PUBLIC KEY ENCRYPTION

The main problem in symmetric key encryption is secure exchange of keys between two entities who want to communicate with one another. Suppose A and B want to communicate with one another. The following methods can be used for exchanging keys:

1. A and B exchange keys via a secure channel other than the one which will be used for business communication. In other words A can send B the key by post or secure fax. The major problem is that if A has 1000 business partners with whom it has to communicate 1000 unique keys have to be sent. They should also be stored in a secure table which has to be looked up before each transaction. Further, if the same key is used over a long period there is a risk of leakage of the key. Thus, the keys used by partners have to periodically changed.
2. A and B can use the services of an intermediary C. A and B can request a key for each transaction and C can send it via a separate secure channel. If C has N clients and any one of them wants to communicate with any of the other (N − 1) clients, the total number of independent keys to be managed by C is $N(N − 1)/2$ which can become very large.

For example, if, $N = 1000$, $N(N - 1)/2 = 4999500$. If by chance the database which C uses to store the keys is illegally accessed by an intruder, it will compromise all its clients. As key distribution and management is cumbersome in symmetric key encryption, there was a search for a method which would eliminate this problem. Till about 1970 all encryption was using secret keys and primarily used the confusion-diffusion idea. A big breakthrough was achieved in 1976 by Diffie and Hellman who proposed what is known as the *public key encryption method*. The basic steps in public key encryption are:

1. Each user generates a pair of keys called a *public key* and a *private key* for encryption and decryption of messages.
2. Each user places his public key in a database accessible to all other users who want to communicate with him. He keeps with himself the companion private key as a secret key.
3. Referring to Figure 5.8 if Ajit wants to send a secret message to Balu, he encrypts the message with Balu's *public key*.
4. When Balu receives the encrypted message from Ajit, he decrypts it using his *private key*. Anyone who is able to access the encrypted message from the communication link will not be able to decode it as he needs Balu's private key to decode which is known only to Balu.
5. If Ajit wants to communicate with several others he can go to their respective public key databases, retrieve them and use them. If he intends to communicate with them frequently, he can store all their public keys in a database in his own computer. Whenever he wants to communicate with one of them, say, Chandru, he can retrieve Chandru's public key and use it.
6. The decryption key used by any person or organization is private and generated locally. Thus, there is no key distribution problem. If an organization has some suspicion that its private key has leaked, it can immediately change its private key and replace its old public key with a new public key corresponding to the new private key.

Another important property of the public key encryption system is that it is reversible. If a message is encrypted with a private key, it can be decrypted with the corresponding public key. This property can be used for message authentication which we will examine later.

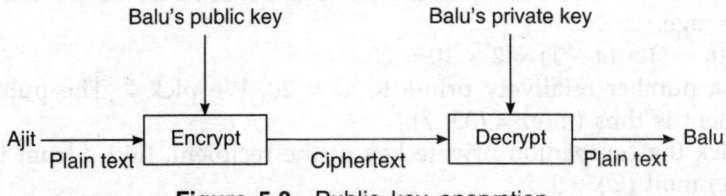

Figure 5.8 Public key encryption.

5.3.1 RSA Encryption Scheme

Currently the most popular public key encryption method was proposed by Rivest, Shamir and Adleman in 1978 and is called the RSA scheme. It is a block cipher in which the plain text and ciphertext and keys are usually 1024 bits long (around 309 decimal digits). It can, however, work with any block size with integers 0 to $(n-1)$ for some n.

The encryption and decryption in RSA is performed as shown below.

Let M = plain text block ($\leq n$) M and n are integers
   C = ciphertext block

Given M

$C = M^e \bmod n$ Encryption scheme e is an integer
$M = C^d \bmod n$ Decryption scheme d is an integer

The operation $M^e \bmod n$ and $C^d \bmod n$ are known as modular exponentiation.

The public key is (n, e) and the private key is (n, d).

Given n we have to find a pair e, d which satisfies the encryption and decryption schemes. This is done by using the following process.

Step 1: Pick two large prime numbers p and q

Step 2: Find $\emptyset = (p-1) \times (q-1)$

Step 3: Find e relatively prime to $\emptyset$, i.e., gcd $(\emptyset, e) = 1; 1 < e < \emptyset$

   (n, e) is the *public key*

Step 4: Find a number d which satisfies the relation $(d \times e) \bmod \emptyset = 1$

   (n, d) is the *private key*

We will give a small example to illustrate the method. In practice, p and q are very large primes which are at least 150 digits long to protect the code from brute force (i.e., exhaustive enumeration) attempts to crack it. The RSA system relies on the fact that given a number which is the product of two prime numbers, it is difficult to factor the product to find the two prime numbers.

Example: We find the public and private keys of a recipient of a ciphertext.

1. Pick prime numbers $p = 3, q = 11, n = p \times q = 33$

   Note: This implies that the block value should be ≤ 33. As we are using decimals, we will use decimal encoding of messages. If we pick 26 uppercase English letters (A to Z $\rightarrow$ 1 to 26) and $(a, b, c, d, e, f \rightarrow$ 27 to 32) and do letter by letter encryption this value of n is sufficient. This set of 32 letters is picked to represent a message.
2. $\emptyset = (p-1) \times (q-1) = 2 \times 10 = 20$
3. Pick a number relatively prime to $\emptyset = 20$. We pick 7. The public key of the recipient is thus $(n, e) = (33, 7)$
4. To pick the companion private key of the recipient, find d from the relation
   $(d \times e) \bmod (\emptyset) = 1$
   $(d \times 7) \bmod 20 = 1$ which gives 3 for d.
   Therefore, the private key of recipient is (33,3).

We will now apply the RSA algorithm to a message: CNDE whose decimal equivalent is 3, 14, 4, 5.

Encryption: Ciphertext (C) = 3^e mod n = 3^7 mod 33
= 2187 mod 33 = 9

Thus, first letter is I.
The second letter encrypted is
$(14)^7$ mod (33) = 105413504 mod 33 = 20.
Thus, the second letter is T.
The third letter encrypted is
$(4)^7$ mod (33) = 16384 mod (33) = 16.
The third letter is P.
The last letter encrypted is
$(5)^7$ mod (33) = 78125 mod (33) = 14.
The last letter is N.

Thus, the ciphertext corresponding to the plain text CNDE is ITPN. Let us now decrypt this ciphertext to plain text using the private key of the recipient of the ciphertext.

Decryption: Ciphertext: ITPN = 9, 20, 16, 14
The private key is (33, 3).
Plain text is derived using the equation M = C^d mod n
Decrypting character by character, we get

I → 9: 9^3 mod 33 = 729 mod 33 = 3 → C
T → 20: 20^3 mod 33 = 8000 mod 33 = 14 → N
P → 16: 16^3 mod 33 = 4096 mod 33 = 4 → D
N → 14: 14^3 mod 33 = 2744 mod 33 = 5 → E

Thus, plain text is CNDE as it should be.

The RSA algorithm is computationally more complex than DES as we have to find high powers of an integer particularly if e and d are large. Even though reasonably efficient algorithms have been proposed to reduce the time it is still substantially larger than DES. Finding large primes p and q is also computationally complex. It is also time consuming to find d and e for a given n. RSA requires large block sizes to ensure security and this in turn makes it substantially slower than DES.

As we saw the security of RSA depends on the difficulty of factoring large number n to its prime components p and q. Many approaches have been proposed but if n is sufficiently large, finding p and q is quite difficult. We mentioned another interesting property of RSA encryption, namely, the algorithm is symmetric. In other words, if a plain text is encrypted with a private key, it can be decrypted with the corresponding public key. This is used in authenticating a message and signing it digitally. We will describe this in a later section. In this section, we will show the symmetric character of the RSA algorithm. Given the public key, private key pair (33,7) and (33,3) we will encrypt a message sent by A with his private key (33,3) and see how the recipient B can decrypt it with A's public key.

Let the message be CNDE whose decimal equivalent is 3, 14, 4, 5.
Remember that C = M^d mod n
Encryption with private key (33,3) of A gives

Ciphertext (C) = 3^3 mod 33 = 27 mod 33 = 27 = a

(A to Z is 1 to 26, and 27, 28, 29, 30, 31, 32 are coded by a, b, c, d, e, f)

Ciphertext of N = 14^3 mod 33 = 2744 mod (33) = 5 = E
Ciphertext of D = 4^3 mod 33 = 64 mod 33 = 31 = e
Ciphertext of E = 5^3 mod 33 = 125 mod 33 = 26 = Z

Thus, the ciphertext is aEeZ.

Decryption with the public key (33, 7) of the letter a gives

$$M = C^e \bmod n$$

Plain text of a = $(27)^7$ mod 33 = 3 = C

Plain text of E = $(5)^7$ mod 33 = 14 = N

Plain text of E = $(31)^7$ mod 33 = 4 = D

Plain text of Z = $(26)^7$ mod 33 = 5 = E

(Remember that $(x)^7$ mod 33 = (x^3 mod 33 × x^3 mod 33 × x mod 33) mod 33 which is easy to calculate).

This example shows that a plain text encrypted with the private key of a sender can be decrypted using the sender's public key. In this simple example, we assume that special procedures to map A to Z, a to f to integers 1 to 32 are used.

5.3.2 Diffie–Hellman Key Exchange Algorithm

There is another public key encryption methods which is important. This is called Diffie–Hellman key exchange algorithm developed by the original proposers of the public key encryption idea. The advantage of this method compared to RSA is the fact that the key to exchange messages by sender and receiver is identical and each can calculate it using the public key exchanged between them. The algorithm depends for its effectiveness on the difficulty of computing discrete logarithms in groups.

The steps of the algorithm are:

Step 1: Two numbers are selected. We will call them q and a. q is a prime number. $a < q$ is a primitive root of q. a is a primitive root of q if the following is true:

a mod q, a^2 mod q, a^3 mod q, ..., a^{q-1} mod q are distinct and are numbers 1 to $q - 1$ in some permutation.

Step 2: The sender selects a random number XS < q which is private to him. The sender calculates his public key YS = a^{XS} mod q.

Step 3: The receiver selects a random number XR < q. XR is private to him. The receiver calculates his public key YR = a^{XR} mod q.

Step 4: The sender and receiver exchange their public keys.

Step 5: S generates secret key K = $(YR)^{XS}$ mod q

Step 6: R generates secret key K = $(YS)^{XR}$ mod q
It can be seen that K = $(YR)^{XS}$ mod q = $(a^{XR})^{XS}$ mod q
= $(YS)^{XR}$ mod q = $(a^{XS})^{XR}$ mod q

In other words, the secret key K of both the receiver and sender are identical and can be calculated from each other's public key and private random number XS and XR.

Message exchange procedure

1. The sender S picks q and a and forwards them to R.
2. The sender S forwards his public key YS to R.
3. The receiver R sends his public key YR to S.
4. Using each other's public key R and S independently compute a secret key K.
5. Now S can send a message to R using 3DES encryption using the key K.
6. R can decrypt the encrypted message using his own copy of K. The DES key need not be exchanged as both the sender and the receiver calculate it locally.

We give a toy example of the key calculation taken from the book by Stallings:

Let $q = 71$. A primitive root a of q is 7
Select randomly XS = 5 and XR = 12
S computes YS = 7^5 mod 71 = 51 as his public key
R computes YR = 7^{12} mod 71 = 4 as his public key
S and R exchange their respective public keys
Common secret key is calculated as
K= $(YR)^{XS}$ mod 71 = 4^5 mod 71 = 30
K = $(YS)^{XR}$ mod 71 = 51^{12} mod 71 = 30

Given the two public keys 51 and 4 an attacker cannot easily find the common secret key 30.

This method ensures both secrecy and authenticity of messages exchanged between the sender and the receiver as the secret key could not have been calculated from the public keys which each of them calculated using q and a. As XS and XR are secret, K cannot be calculated even if q and a are public knowledge. Thus, q and a can be sent to R by S along with YS.

It has been found that if q, a and YR are captured by an attacker he can try various values of XS and XR and calculate K. Thus, in this method it is safer to change the public keys in each *session* (or transaction) between the sender and receiver. The Diffie–Hellman key exchange method is used in B2C e-commerce. We give one example below of a customer and a merchant negotiating sale of a product with the customer using a credit card. This is a very simple protocol in which the merchant is able to read the customer's credit card number along with the purchase order. The sequence of exchange of information between the client computer (customer) and the server computer (merchant) is explained below (See Figure 5.9).

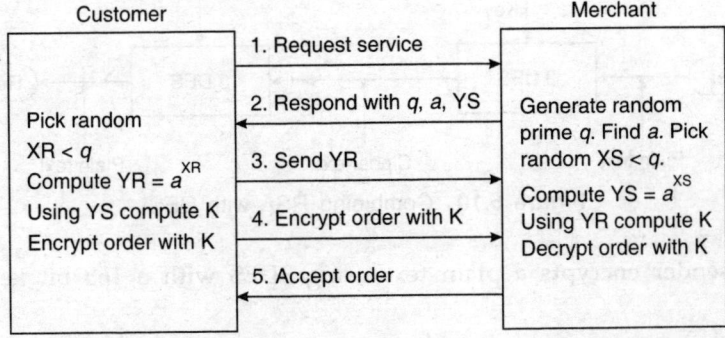

Figure 5.9 Diffie–Hellman public key exchange algorithm.

Step 1: A customer logs on to the site of the merchant requesting service.

Step 2: The merchant's server selects a random prime number q and calculates a the primitive root of q. The server selects a random number $XS < q$ and calculates $YS = a^{XS}$ mod q. YS is the merchant's public key which is sent to the customer along with q and a by the merchant's server.

Step 3: The customer's computer selects a random number $XR < q$. It calculates its public key $YR = a^{XR}$ mod q and sends it to the merchant's server.

Step 4: The customer's computer generates a secret key $K = (YS)^{XR}$ mod q. It encrypts the credit card number and purchase order with K and sends it to the merchant's server.

Step 5: The merchant's server locally calculates K as $(YR)^{XS}$ mod q and decrypts the credit card number and purchase order. It sends an acknowledgement to the customer.

Observe that a customer need not have his or her own public key. It is generated for each transaction "on the fly" using random values of q and a generated by the merchant's server and locally generated private XS. Also observe that the secret key is valid only for one transaction. Thus, it cannot be broken by an intruder.

In practice, the customer need not know any algorithm. Most browsers will implement this secure transaction when requested.

5.3.3 Combining RSA with DES

We saw that 3DES is computationally simple to implement and fast whereas RSA is computationally complex. The primary disadvantage of 3DES is the key distribution problem which is simpler in the RSA system. If we can combine the two systems to get the advantage of computational simplicity of 3DES without having the problem of key distribution it will be a good system. This can be done as shown in Figure 5.10. The steps followed are:

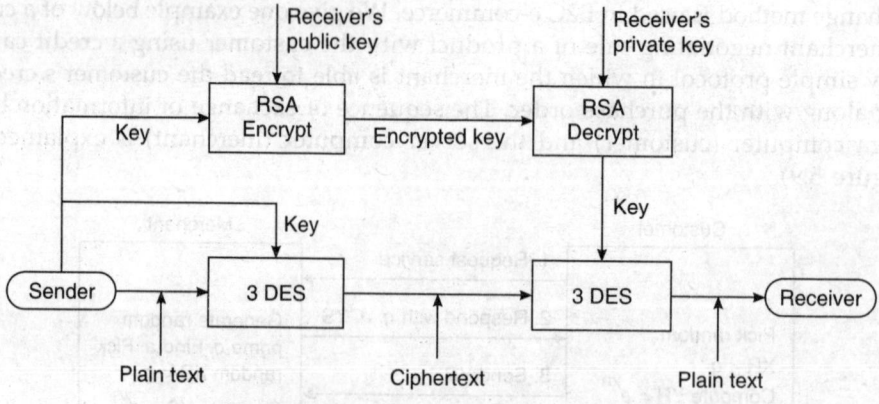

Figure 5.10 Combining RSA with DES.

Step 1: The sender encrypts a plain text using 3DES with a 168 bit key obtaining a ciphertext.

Step 2: The DES key is encrypted with the receiver's public key using the RSA encryption algorithm.

Step 3: The receiver decrypts the DES key with his private key using the RSA algorithm.

Step 4: Having found the DES key the receiver decrypts the ciphertext and obtains the plain text.

Step 5: The sender can use a different key with each plain text he sends ensuring that an eavesdropper cannot decrypt the ciphertext even if he is able to access several ciphertexts. (Each transmission of a plain text is called a session).

The main advantages of this method are:

1. As the DES key is small, encrypting it using RSA is fast. The plain text can be very long and 3DES will encrypt it much faster than RSA.
2. The DES key is changed in each session making it almost impossible to decrypt the message even if several encrypted sample messages are accessed by a hacker.
3. DES key distribution problem is eliminated. However, RSA public key distribution problem still remains. This is solved by a public key distribution and authenticating organization which we will discuss in the next section.

5.4 PUBLIC KEY CERTIFYING AUTHORITY

In RSA an entity A has to know the public key of another entity B to be able to send encrypted information to it. If A sends an e-mail to B and requests its public key it may be intercepted by a third party C which may send its public key to A. A will not know this and all communications from A to B henceforth can be decrypted by C. Thus, A must be sure that the public key it gets from B does indeed belongs to B. Further, A must be sure that B is a legitimate business and can be trusted. To ensure this most countries have established a hierarchical system of public key certifying authorities which issues a public key certificate to an individual or organization for a fee. In India, the Government has delegated the right to issue public key certificates to some agencies like NIC (National Informatics Centre) CCA-MCIT (Ministry of Communications and Information Technology), and to some companies such as Tata Consultancy Services Ltd., CMC Ltd., etc., as public key certifying authorities. The procedure followed is described below:

1. An entity wanting its public key certified usually sends by registered post or courier its public key to be certified, certified copy of postal address and a certificate such as sales tax certificate or income tax certificate or income tax PAN number. These documents are needed by the authority to ensure that it is a genuine company. An individual would normally send along with his/her public key, address proof such as telephone bill and passport copy certified by a notary by registered post. Certifying authority may also accept these documents encrypted using the authority's public key.
2. The certifying authority may in case of doubt send inspectors to verify the information given by the applicant. Else, it will certify the public key by attaching the approved public key certificate. The certificate has an internationally agreed upon format as shown in Table 5.2(a).

Table 5.2 Standard formats of digital certificate and revocation list

Version	Signature algorithm identifier
Certificate serial number	Issuing authority name
Signature algorithm identifier	This update date
Name of issuing authority	Next update date
Issue date and validity period	Revoked certificate serial number
Organization's name	Revocation date
Organization's public key	Organization's name
Authority's unique identifier	Authority's unique identifier
Authority's signature	Authority's signature
(a) Format of public key certificate	(b) Format of certificate Revocation list

It has information such as name of the organization, a unique identification number and time stamp of when the certificate was issued and its validity period.

1. Suppose A wants its public key certificate. It applies to the certifying authority with all relevant documents. If the authority is satisfied, it will approve the certificate. It will be encrypted with the authority's private key and sent to A. It also maintains a list of public keys which are revoked [See Table 5.2(b) for the format]. We explain later why a public certificate may be revoked.
2. This certificate can be decrypted using the public key of the certifying authority. This certificate is an important document which A can intimate to any other entity which wants to have e-commerce transactions with A.
3. Suppose B wants to transact business with A, the following procedure is followed by B (See also Figure 5.11). B requests A for its public key certificate. A sends to B the certificate encrypted with the certifying authority's private key which it

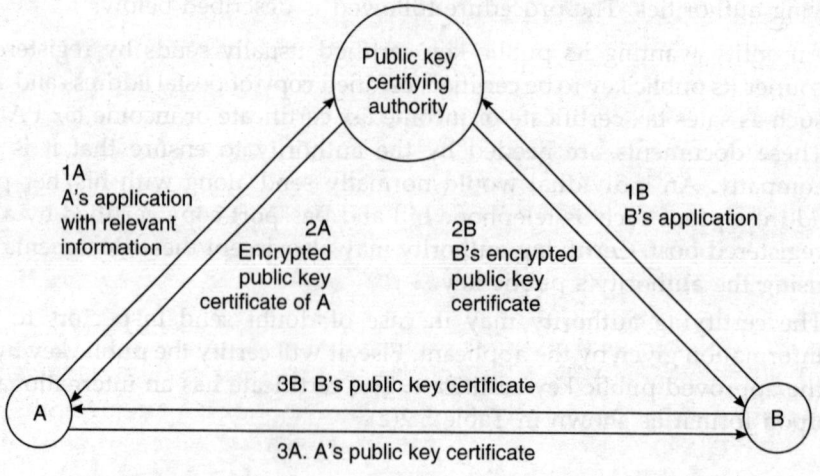

Figure 5.11 Obtaining public key from certifying authority.

had received in Step 1. B can decrypt it using the public key of the certifying authority. The fact that the certificate is encrypted with certifying authority's private key gives B the assurance that the public key of A is indeed given by the certifying authority and is not a fake. From the decrypted certificate B can extract A's public key and other details such as validity period of the certificate and whether it is currently valid. B is now confident that it can transact business with A. Before this B should also send its public key certificate to A following a procedure similar to the one followed by A.

A public key may be revoked by a certifying authority for several reasons such as bankruptcy of a company or on the request of the company if it doubts that its private key is compromised. Thus, the authority periodically publishes a revocation list whose format is given in Table 5.2(b). Thus, a business should access the publicly accessible revocation list of the certifying authority before using the public key of any business with whom it wants to transact business.

In India there is one more level called controller of public key certifying authorities. The controller maintains a database of certificates issued by all certifying authorities approved by it. The controller is the legal authority for dispute resolution.

5.5 DIGITAL SIGNATURE

So far we have described the methods of encrypting documents exchanged between entities participating in e-commerce. There is another important requirement when documents are exchanged using an electronic communication medium. When a document such as a purchase order is sent by post, it has two important characteristics which has to be imitated by electronic documents. First, the letter head of a business, its seal and the signature of a person in the document convinces the recipient of the authenticity of the document. Second, the signature physically appears following the text and this ties the signature to the typed matter in the document. In a legal document, every page is signed and every correction is also signed. We have to mimic the physical signature with an electronic equivalent. There are two important requirements to be met by a digitally signed electronic document.

1. A receiver R receiving a document from a sender S has to be sure it is from S.
2. The signature should be tied to the document sent by S. This will ensure that S cannot later claim that he or she never sent the document to R, in other words S cannot repudiate his or her communication.

These requirements are met by the following scheme (See also Figure 5.12).

Step 1: Before starting a transaction, S and R exchange their public key certificates issued by a certifying authority. This is necessary to meet condition 1 mentioned earlier.

Step 2: S encrypts the document D to be sent to R with R's public key getting ciphertext DE.

Step 3: R decrypts DE using its own private key and gets the original document D.

Step 4: S makes a unique abstract of the document D called the hash of D which we will represent by H(D). (The requirements of a function H to give a unique hash of D will be described later). H(D) is encrypted by S with its *private key* getting H(D)E.

Step 5: R decrypts H(D) E using the public key of S getting back H(D).

Step 6: Using D obtained in step 3, R creates the hash of D, namely, H(D) using the known hash function H.

Step 7: R compares H(D) obtained in step 5 with that obtained in step 6. They should exactly match. If they do, then R accepts the document as authentic document signed by S. Else, it is rejected as a forgery.

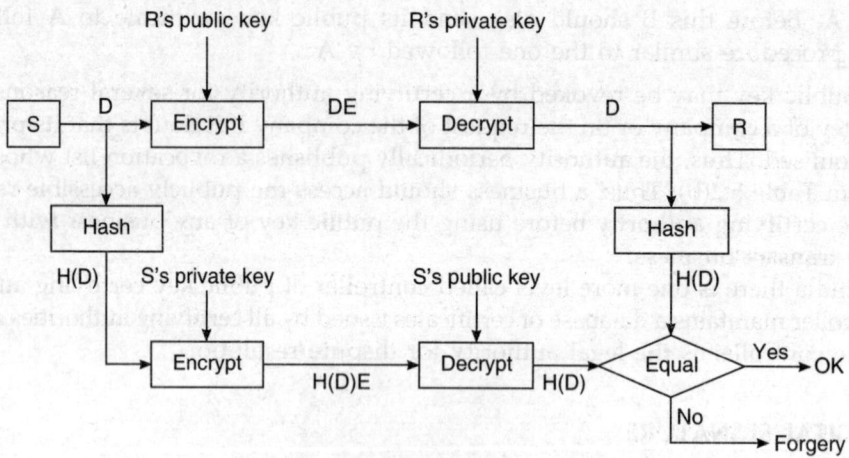

Figure 5.12 Digitally signing a document.

We have used the term hash function in the above description. The hash function is a function which takes a text and replaces it by an abstract which is usually much shorter. For example, one method of hashing is to take the ASCII value of each character, add these and get a sum N. This hashing is not the one used. A standard algorithm known as MD5 algorithm is used. This algorithm guarantees that two different texts D1 and D2 will give two different hashes. In other words, if D1 ≠ D2, H(D1) ≠ H(D2). It is also necessary for a hash function, that given a hashed value x we cannot find a y easily which will make H(y) = x. In other words one cannot "manufacture" a document given its hash value. Mathematically H is called a one way function. The fact that H(D) is almost always unique for a given D and that H(D) is encrypted by S using his or her *private key* which is known only to him or her assures R that only S could have sent the document D. It also assures R that the document received by him could not have been tampered by any hacker as H(D) created by R and that sent by S encrypted with his private key and decrypted by R match.

In the previous sections, we described a few of the encryption algorithms. There are several others we have not described. In Table 5.3, we give a list of some other algorithms.

Table 5.3 Security related algorithms

Single key encryption	Public key encryption	Hash functions
DES–56 bit key	RSA	MD5 (128 bit digest)
3 DES–168 bit key	Diffie-Hellman	SHA-1 (160 bit digest)
IDEA–128 bit key	Elliptic Curve	RIP MED (160 bit digest)
AES–128, 192, 256 bit key		DSS
RC2, RC4, RC5 (40 to 128 bit keys)		

We will briefly give some information of the other systems in what follows:
- IDEA (International Data Encryption Algorithm) is a symmetric key encryption similar to DES. It also uses permutations and substitutions. Its details are different from DES. It is considered more difficult to break compared to DES.
- RC2, RC4, RC5 are developed by RSA systems company (Details may be found in Stallings [11].
- Elliptic curve public key encryption method is considered mathematically sound and computationally less intensive compared to RSA. Products are just emerging using this algorithm.
- SHA-1 (Secure Hash Function) was developed by National Institute of Standards and Technology (NIST), USA and is a US Govt. standard. The algorithm takes an input message of length $\leq 2^{64}$ bits and produces a 160-bit message digest. The input is taken as 512-bit blocks and after all blocks are processed we get 160-bit digest. It satisfies all the requirements of an ideal hash function.
- RIPMED is A European Standard similar to SHA-1.
- DSS Digital Signature Standard is proposed by NIST. It uses SHA1 message digest and private key for signing.

5.6 SOME APPLICATIONS OF ENCRYPTION TECHNOLOGIES

5.6.1 Secure E-mail

Electronic-mail is now universally used for business communications. In early days mail was primarily text and a protocol called SMTP (Simple Mail Transfer Protocol) was used to transmit mail from a sender to a receiver on the Internet. Nowadays e-mails include text, images, audio, video, spreadsheets, powerpoint transparencies, etc. Thus, a new mailing protocol called MIME (Multipurpose Internet Mail Extension) has been standardized and is now widely used. This protocol does not provide any security to e-mail correspondence. In other words, any intruder who is able to access the net can read the mail and may even be able to alter it. Thus, there is a requirement for businesses to send secure e-mail. The primary requirements of a secure e-mail system are confidentiality and authentication. The requirement of confidentiality is met by encryption and that of authentication by digital signature. This has been incorporated in a new system called S/MIME (Secure MIME). Thus, cooperating businesses must both install S/MIME software in their PCs connected to their intranets. The Internet infrastructure will then be able to transport the message from sender to a receiver. S/MIME has been standardized and public domain software to implement it is available. Besides this, companies such as RSA Security have products to implement S/MIME.

The method adopted by S/MIME to ensure confidentially is to use AES, 3DES or RC2 symmetric encryption for e-mail. For each e-mail a random key is used. The sender delivers this key to the receiver using RSA encryption. For authentication the e-mail is digitally signed by using MD5 or SHA-1 digest and encrypting using RSA private key. Digital certificates issued by a certifying authority are exchanged between businesses to authenticate their respective public keys.

5.6.2 Secure Socket Layer

Netscape communications originally proposed a protocol layer between TCP/IP protocol stack and application layer (Figure 5.13). This is called *Secure Socket Layer* (SSL) whose purpose is to provide end-to-end security for applications such as e-commerce using the web. It was adopted by several browsers such as Internet Explorer and Firefox.

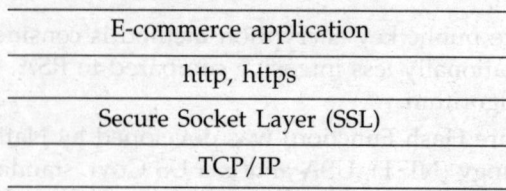

Figure 5.13 Security layer to provide application level security.

Currently it has been standardized as Transport Security Layer (TSL). TSL secures the TCP/IP connection by using one of several optional encryption methods, digital signature software and public key certificates. The protocol treats a client and a server as equals (peers) as far as requests, replies and their security preferences are concerned. Usually, when a client which has a browser with SSL logs on to a server, the server acknowledges and sends back normally its public key certificate. This authenticates the legitimacy of the e-commerce site of the server. In B2B e-commerce, SSL ensures that the client and server exchange public key certificates, security methods to be used for encryption and digital signature standard. After this transactions proceed. However, in B2C e-commerce the customer does not have a public key. Thus, the business does not insist on this certificate. The client's order including the credit card number is encrypted with the public key of the server and sent. Usually the RSA public key is used. The server decrypts the client's order with its private key. Usually in the server-client negotiation phase, the server's public key certificate is requested (by the browser) from which the client is able to get assurance of the vendor's legitimacy. Another alternative provided is for the client to pick a random symmetric key for one transaction, encrypt with merchant's public key and send it to the merchant's server. After this all transactions take place using a symmetric key algorithm such as Triple DES.

The advantages of SSL are:

- It is supported by almost all browsers.
- End-to-end security of transaction is provided, that is, multiple encryption need not be done at any intermediate points.
- A range of security options are provided depending on the negotiation between the client and the server.

Among security methods supported by SSL, the Diffie–Hellman public key exchange algorithm is also one. Use of this method follows the scheme explained in section 5.3.2 (See Figure 5.8).

5.6.3 Secure Hypertext Transfer Protocol

The normal hypertext transfer protocol is not secure. In order to provide security to http transactions browsers with SSL support secure http which is labeled https. If a customer wants to transact business with a merchant he or she should use https. If the merchant's server or a customer's client does not support SSL, then the https request will be rejected with a message "cannot access using this protocol". Nowadays all clients and all servers implement SSL security. Thus, https is universally supported. The roles of https and SSL are different. https is an application and depends on SSL infrastructure to ensure end-to-end security of transactions in e-commerce. http and https can exist simultaneously in a browser. When secure transaction is needed https is invoked, else http is used.

5.7 CONCLUSIONS

We have described various schemes for ensuring security of e-commerce transactions. The web server of an organization should be protected from hackers while allowing legitimate users to access it. We have seen that firewalls and IP-Sec enabled routers prevent unauthorized entry to corporate intranets. The web site of an organization is intended to publicize the organization and it must be accessible to all. Information in such a site should be partitioned into two parts—one publicly accessible and the other confidential. For instance, in an e-commerce site if a customer enters his or her credit card number it is stored temporarily. This data should be encrypted before it is stored. All sensitive data should be encrypted using an encryption method such as 3 DES (with the key known only to a system administrator) and stored. It is also advisable to take a back up, for instance, in a CDROM and store it off-line and purge the hard disk. This will reduce the risk of compromising the customers' credit card numbers if stored on-line longer than needed. Many organizations have a security policy in which Internet connectivity is not provided to intranets where sensitive information is stored. Sensitive information would normally include company's strategies, personal information, software systems and similar valuable data.

All the above security methods are intended to prevent non-company individuals from accessing sensitive information. Very often the enemy may be within an organization. Thus, organizations must have in place an overall company security policy which should be comprehensive. Several books discuss these issues in great detail. They are outside the scope of this book.

SUMMARY

1. Message exchanged between organizations using the Internet can be easily tapped by eavesdroppers. It is thus necessary to scramble them to prevent eavesdroppers from understanding the messages. It is done by encrypting messages.
2. Message (plain text) is encrypted by substitution and transposition. The encrypted text is called ciphertext.
3. Substitution is done either by replacing a character by another character a number of places away from it in the collating sequence or by table lookup. Transposition is usually done by permutation of characters by a specified permutation operator.
4. This general idea is used in a standard encryption method called Digital Encryption Standard (DES). DES encrypts 64-bit blocks with a 56-bit key.

5. With faster computers DES key can be broken by exhaustive search. It is not secure and has been replaced by Triple DES (3 DES) which uses DES algorithm thrice with three different 56-bit keys. 3 DES is quite secure.
6. DES has been implemented as a hardware device. 3DES hardware may be attached to a computer's output port so that messages sent from the computer are encrypted. The receiver can decrypt it if the key is given.
7. In all encryption methods, algorithm is public knowledge. The secrecy lies in the key.
8. A system in which the encrypting and decrypting keys are the same is called a symmetric key system. 3DES is a symmetric key system.
9. Another symmetric key encryption method was standardized in 2000. It is called Advanced Encryption Standard (AES). It encrypts 128-bit blocks using keys of length 128, 192 or 256 bits depending on the level of security needed.
10. AES algorithm is publicly available and non-proprietary. It can also be implemented as hardware.
11. The main problem with a symmetric key system is the need to distribute the key securely to all participating businesses. Symmetric key encryption/decryption is fast.
12. Two-key based system called RSA system does not require distributing secret keys. It has two keys for each participant in the communication, a private key and a public key. If A wants to send a message to B, A encrypts the message using B's public key. B decrypts it using his private key. Thus, there is no key distribution problem. It is, however, slower than the symmetric key system.
13. RSA system is based on the fact that it is difficult to factor two prime components from their product, particularly, when the prime numbers are large.
14. In RSA system, a message encrypted with a private key, can be decrypted with the corresponding public key. This is used in digital signature.
15. In order to sign a message the sender hashes the message with a known algorithm to get a message digest MD. MD is encrypted with a sender's private key and sent to the intended receiver. Let us call it MD_e. The message itself is encrypted with a symmetric key and sent. The recipient decrypts the message and computes the message digest MD using the known hashing algorithm. The recipient then decrypts the encrypted message digest MD_e using the sender's public key. If $(MD)_e$ decrypted = MD then the message is not a forgery as only the sender knows his or her private key. This signature is tied to the message and cannot be repudiated by the sender. The hash function must have the property $H(x_1) \neq H(x_2)$ where $x_1 \neq x_2$ and given $H(x) = y$ we cannot find x given y.
16. To ensure that public keys of organizations do belong to them there are certification authorities which check the legitimacy of organizations and issue public key certificates.
17. There is another public key algorithm called Diffie–Hellman key exchange algorithm. The main advantage of this method compared to RSA is that it uses a symmetric key employing public keys exchanged between sender and receiver. The symmetric key value need not be exchanged.

18. Encryption of multimedia e-mails use what is known as Multipurpose Internet Mail Extension (MIME). Security-enhanced MIME is called S/MIME. S/MIME uses 3DES to encrypt messages. The session key is sent by the sender encrypted using the receiver's public key using RSA encryption. In B2B e-commerce the two parties exchange their public key certificates. If e-mail is to be signed, it uses the MD5 hashing function and hashed message is encrypted with the sender's private key.
19. SSL (Secure Socket Layer) which is a protocol layer between TCP/IP layer and application layer is used by several browsers to provide end to end security of e-commerce transactions. It also uses a combination of RSA and symmetric key cryptography.
20. Browsers which support SSL also support security enhanced http called https to access web sites and send information such as credit card numbers securely over the Internet.

EXERCISES

5.1 Enumerate the threats to the security of a web site maintained by an organization.
5.2 When a message is sent by A to B on the Internet what are the types of security problems which may be encountered?
5.3 Distinguish between confidentiality, authenticity and non-repudiation of messages sent between a customer and a company in e-commerce.
5.4 Why is encryption necessary when a message is sent using the Internet?
5.5 Explain what you understand by symmetric key encryption. Why is it used?
5.6 Define the terms plain text, ciphertext, encryption, decryption, cryptanalysis, substitution and permutation used in cryptography. Given a plain text "WEWILLMEETONSIXTH" encrypt it using the transposition A → F, B → G .... and permutation (3 2 1 5 4). Explain how it is decrypted.
5.7 What is Digital Encryption Standard? How many bit blocks does it use for encryption? What is the key length? What operations does it use for encrypting a plain text?
5.8 Describe the DES algorithm briefly giving the various steps it uses for encryption. Is it considered secure currently? If not, explain why it is not secure.
5.9 What is the difference, if any, between encryption and decryption algorithms in DES? Is it possible to implement DES encryption in hardware? If yes, explain whether the same hardware chip can be used to encrypt and decrypt.
5.10 What is Triple DES (or 3DES)? Explain with a block diagram how Triple DES is implemented using DES encryption blocks.
5.11 What is the size of blocks used in 3DES? What is the size of keys used in 3DES? How many rounds of transcription and permutation are applied in 3DES algorithm? Is it more secure than DES?
5.12 What were the important conditions specified by National Institute of Standards Technology for a new standard symmetric key encryption system?

5.13 What is Advanced Encryption Standard (AES)? What are the block and key sizes used in AES?

5.14 Give the gist of AES algorithm.

5.15 What is the difference between symmetric and public key system of cryptography? Explain how a public key system is used. List the advantages and disadvantages of a public key system when compared to symmetric key encryption system.

5.16 Is a public key system reversible? (i.e., if message is encrypted with a private key can it be decrypted with the corresponding public key). If yes, what is the advantage of this?

5.17 Explain RSA algorithm. Illustrate how the encryption and decryption works using two prime numbers 5 and 7. What are the public and private keys of the system?

5.18 What is the normal key length of RSA system used to ensure security?

5.19 Explain Diffie–Hellman key exchange algorithm. In what way is it different from RSA algorithm? Compare and contract RSA and Diffie–Hellman algorithms.

5.20 Explain how the Diffie–Hellman key exchange algorithm can be used by a customer and a vendor for secure e-commerce transaction.

5.21 Can RSA and DES be combined? If yes, how is it done? What is the advantage, if any, of doing it?

5.22 What is a public key certifying authority? Why is such an authority needed? What are the services provided by a public key certifying authority?

5.23 What is the procedure used by a public key certifying authority to issue a public key certificate? What information is included in such a certificate?

5.24 What is a digital signature? How is an e-document signed by a sender and sent to a receiver? How does the recipient of a signed e-document authenticate the signature?

5.25 What is a hash function? What are the conditions to be fulfilled by a hash function in order that it can be used in a digital signature application?

5.26 What is S/MIME? Where is it used? How does it ensure confidentiality and authenticity of messages?

5.27 What is SSL? Is it different from TLS?

5.28 What is the role of SSL in e-commerce?

5.29 If a credit card transaction is to take place between a customer and a merchant what is the role of SSL and what is the role of http?

5.30 Does SSL allow absence of client level public key and public key certificate?

5.31 Explain step by step how a client can use SSL and https to order an item (or list of items) from a vendor and ensure secure payment using his credit card.

5.32 What is secure hyper text transfer protocol? Briefly explain how it functions.

5.33 What are the methods used to ensure the security of the web site of an organization?

5.34 Why is it required to encrypt customer data in an organization's web site? What is the method of encryption used for this?

5.35 Is it advisable to isolate the intranet of an organization from the Internet? Under what circumstances would one advise this?

OBJECTIVE QUESTIONS

Each question has four possible answers. Pick the most appropriate answer.

5.1 By encryption of a message, we mean
   (a) Compressing it
   (b) Expanding it
   (c) Scrambling it to preserve its secrecy
   (d) Hashing it to a shorter abstract

5.2 Encryption is required to
   (i) Protect business information from eavesdropping when it is transmitted on the Internet
   (ii) Efficiently use the bandwidth available in PSTN
   (iii) Protect information stored in companies' databases from retrieval
   (iv) Preserve secrecy of information stored in databases if an unauthorized person retrieves it
      (a) (i) and (ii)   (b) (ii) and (iii)
      (c) (iii) and (iv)   (d) (i) and (iv)

5.3 Encryption can be done
   (a) Only on textual data
   (b) Only on ASCII coded data
   (c) On any bit string
   (d) Only on mnemonic data

5.4 By applying permutation (31254) and substitution by 5 characters away from current character (A → F, B → G etc.) the following string ABRACADABRA becomes
   (a) FGWCAAADRBF
   (b) RABCAAADRBF
   (c) WFGHFFFIWGF
   (d) None of the above

5.5 The ciphertext PDLJDLXHVQC was received. The plain text was permuted using permutation (34152) and substitution. Substitute character by character +3 (A → D, etc.). The plain text after decryption of the ciphertext is:
   (a) MAIGAIUESNZ
   (b) IAMAGENIUSZ
   (c) LDPDJHPLXVZ
   (d) IAMAGENIUSC

5.6 By symmetric key encryption, we mean
   (a) One private key is used for encryption and decryption
   (b) Private and public key used are symmetric
   (c) Only public keys are used for encryption
   (d) Only symmetric key is used for encryption

5.7 In encryption methods used, the following is true:
   (i) Encryption algorithm is public knowledge
   (ii) Encryption algorithm is secret
   (iii) Encryption key is secret
   (iv) Encryption key is public knowledge
      (a) (i) and (iii)   (b) (ii) and (iii)
      (c) (i) and (iv)   (d) (ii) and (iv)

5.8 The acronym DES stands for
   (a) Digital Evaluation System
   (b) Digital Encryption Standard
   (c) Digital Encryption System
   (d) Double Encryption Standard

5.9 DES works by using
   (a) Permutations and substitutions on 64-bit blocks of plain text using a 56-bit key
   (b) Only permutations on blocks of 128-bits
   (c) Exclusive ORing key bits with 64-bit blocks
   (d) 4 rounds of substitution on 64-bit blocks with 56-bit keys

5.10 DES
   (i) Is a symmetric key encryption method
   (ii) Guarantees absolute security
   (iii) Is implementable as hardware VLSI chip
   (iv) Is a public key encryption method
   (a) (i) and (ii) (b) (ii) and (iii)
   (c) (i) and (iii) (d) (iii) and (iv)

5.11 DES encrypts blocks of 64-bit plaintext using a 56-bit key using
   (a) 8 rounds of permutations with 8 keys
   (b) 16 rounds of substitution and permutation with 16 keys derived from the original 56 bit key
   (c) 32 rounds of substitutions interchanging left and right halves of the 64 block
   (d) 12 rounds of substitutions and permutation with the same key

5.12 Triple DES (3DES) uses
   (a) Modified DES algorithm with a 168-bit key
   (b) DES algorithm thrice on 192-bit blocks with a 168-bit key
   (c) DES algorithm on 128-bit blocks thrice
   (d) DES algorithm on 64-bit blocks thrice each time with a different 56-bit key

5.13 Triple DES compared with DES is
   (a) Three times more secure
   (b) Around 10^{30} times more secure
   (c) Much more secure
   (d) Three times more secure and the algorithm is thrice as efficient

5.14 In the brute force method of breaking DES x keys need to be tried on ciphertext samples to guess the correct key
   (a) $x = 2^{56}$ (b) $x = 2^{64}$
   (c) $x = 2^{128}$ (d) $x = 2^{168}$

5.15 In the brute force method of breaking 3DES x keys need to be tried on ciphertext samples to guess the correct key
   (a) $x = 2^{56}$ (b) $x = 2^{192}$
   (c) $x = 2^{168}$ (d) $x = 2^{64}$

5.16 Advanced Encryption Standard (AES) has the following features:
   (i) Uses 128 bit blocks
   (ii) Key lengths can be 128, 192 or 256 bits
   (iii) Is a symmetric key algorithm
   (iv) The algorithm is in the public domain
   (a) (i), (ii) (b) (i), (ii), (iii)
   (c) (i), (iii), (iv) (d) (i), (ii), (iii), (iv)

5.17 AES uses
   (i) Different algorithms for encryption and decryption
   (ii) Substitution and permutation as the basic encryption ideas
   (iii) 256-bit blocks
   (iv) Minimum 128-bit key
   (a) (i), (ii), (iv)
   (b) (i), (ii), (iii)
   (c) (i), (ii), (iii), (iv)
   (d) (i), (iii), (iv)

5.18 The major problem with all symmetric key encryption systems are
   (a) It is not secure as computers have become very fast
   (b) It is difficult to exchange private keys with several parties
   (c) The algorithms are very complex and time-consuming to implement
   (d) It is not standardized

5.19 Public key encryption method is a system
   (a) Which uses a set of public keys one for each participant in e-commerce
   (b) In which each person who wants to communicate has two keys; a private key known to him or her only and a public key which is publicized to enable others to send messages to him or her
   (c) Which uses the RSA coding system
   (d) Which is standard for use in e-commerce

5.20 Public key system is useful because
   (a) It uses two keys
   (b) There is no key distribution problem as public key can be kept in a commonly accessible database
   (c) Private key can be kept secret
   (d) It is a symmetric key system

5.21 In public key encryption if A wants to send an encrypted message to B
   (a) A encrypts message using his or her private key
   (b) A encrypts message using B's private key
   (c) A encrypts message using B's public key
   (d) A encrypts message using his or her public key

5.22 In public key encryption system, if Alice encrypts a message using her private key
   (a) If a receiver knows it is from Alice he or she can decrypt it using Alice's public key
   (b) Even if a receiver knows who sent the message it cannot be decrypted
   (c) It cannot be decrypted at all as no one knows Alice's private key
   (d) Alice should send her public key with the message

5.23 Message can be sent more securely using Triple DES by
   (a) Encrypting plain text by a different randomly selected key for each transmission
   (b) Encrypting plain text by a different random key for each message transmission and sending the appropriate key to the receiver using a public key system
   (c) Using an algorithm to implement Triple DES instead of using hardware
   (d) Designing Triple DES with high security and not publicizing the algorithm used by it

5.24 In RSA public key system the public key is a tuple (n, e) where
  (i) n is a product of two large prime numbers
  (ii) e is relatively prime to $[(p - 1) \times (q - 1)] < n$
  (iii) n is a large prime number
  (iv) e is a large prime number $< n$
   (a) (ii), (iii)   (b) (iii), (iv)
   (c) (i), (iv)   (d) (i), (ii)

5.25 In RSA public key system the private key
  (i) Is known only to its owner
  (ii) Can be computed if the public key is known
  (iii) Is (n, d) which can be computed if the prime factors of n and the public key are known
  (iv) Can be computed if n is known
   (a) (i), (ii)   (b) (i), (iv)
   (c) (i), (iii)   (d) (ii), (iv)

5.26 The Diffie–Hellman algorithm is
  (a) A private key–public key system similar to RSA system
  (b) More secure than RSA system
  (c) A key exchange method when two parties exchange public keys from which they calculate their own private keys
  (d) A key exchange method when two parties exchange their public key from which they locally calculate a common secret key to exchange messages

5.27 The Diffie-Hellman algorithm is useful
  (a) If a sender does not have a public key
  (b) If a receiver does not have a public key
  (c) When a secret key is calculated locally for each transaction based on private keys exchanged between sender and receiver
  (d) When public keys are exchanged only once and a secret key calculated is used for all future transactions

5.28 The Diffie–Hellman algorithm is useful in B2C e-commerce because
  (a) It is easy to implement
  (b) A customer does not know a merchant's public key
  (c) A secret key common to customer and merchant can be calculated for each transaction even when a customer does not have a public key
  (d) All customers do not have a public key private key pair

5.29 The Diffie–Hellman algorithm ensures
  (a) Confidentiality of data sent by a sender
  (b) Authenticity of data sent by a sender
  (c) Both confidentiality and authenticity of data sent by a sender
  (d) Non-repudiation of data sent by a sender

5.30 Triple DES and public key algorithm are combined
  (i) To speed up encrypted message transmission
  (ii) To ensure higher security by using different key for each transmission
  (iii) As a combination is always better than an individual system

(iv) As it is required in e-commerce
- (a) (i), (ii)
- (b) (ii), (iii)
- (c) (iii) and (iv)
- (d) (i) and (iv)

5.31 A digital signature is
- (a) A bit string giving identity of a correspondent
- (b) A unique identification of a sender
- (c) An encrypted signature of a sender
- (d) An authentication of an e-document by tying it uniquely to a key only a sender can possess

5.32 A digital signature is required
- (i) To tie an electronic message to the sender's identity
- (ii) For non-repudiation of communication by a sender
- (iii) To prove that a message was sent by the sender in a court of law
- (iv) In all e-mail transactions
    - (a) (i) and (ii)
    - (b) (i), (ii), (iii)
    - (c) (i), (ii), (iii), (iv)
    - (d) (ii), (iii), (iv)

5.33 A hashing function for digital signature
- (i) Must be shorter than the message which is hashed
- (ii) Must be hardware implementable
- (iii) Two different messages should not give the same hashed result
- (iv) Is not essential for implementing digital signature
    - (a) (i), (ii)
    - (b) (ii), (iii)
    - (c) (i), (iii)
    - (d) (iii), (iv)

5.34 Hashed message is signed by a sender using
- (a) His or her public key
- (b) His or her private key
- (c) Receiver's public key
- (d) Receiver's private key

5.35 While sending a signed message, a sender
- (a) Sends the message using 3DES and encrypts the hashed message using its private key
- (b) Sends the message using public key encryption and hashed message using 3DES
- (c) Sends both message and hashed message using 3DES
- (d) Sends both message and hashed message encrypted using its public key

5.36 The responsibility of a certification authority for digital signature is to authenticate the
- (a) Hash function used
- (b) Private keys of subscribers
- (c) Public keys of subscribers
- (d) Key used in DES

5.37 Certification of digital signature by an independent authority is needed because
- (a) It is safe
- (b) It gives confidence to a business
- (c) The authority checks and assures customers that the public key indeed belongs to the business which claims its ownership
- (d) Private key claimed by a sender may not be actually be his

5.38 S/MIME is used to send
  (a) E-mail between businesses
  (b) Multimedia securely between businesses
  (c) E-mail between cooperating businesses ensuring confidentiality and authenticity
  (d) Secure e-mail

5.39 S/MIME system
  (i) Uses digital signature with MD5 to ensure authenticity
  (ii) Combines RSA and DES to ensure confidentiality
  (iii) Uses digital certificates for non-repudiation
  (iv) Always uses Diffie-Hellman method for key exchange
    (a) (i), (ii), (iii), (iv)   (b) (i), (ii)
    (c) (i), (ii)   (d) (i), (ii), (iii)

5.40 https
  (i) Is secure hypertext transfer protocol supported by most browsers
  (ii) Is used in B2C e-commerce to enable customer to encrypt his/her credit card number
  (iii) Uses customer's private key for encryption
  (iv) Uses merchant's public key for encryption
    (a) (i), (ii)   (b) (i), (ii), (iii)
    (c) (i), (ii), (iv)   (d) (ii), (iii)

5.41 https normally requires
  (a) A customer's public key certificate
  (b) A merchant's public key certificate
  (c) Any Internet connection
  (d) A virtual private network

5.42 When https is used in B2C e-commerce
  (a) Customer's credit card number is not available to the merchant
  (b) Customer's credit card number is exposed to the merchant
  (c) As the customer encrypts his or her credit card with his or her public key it is confidential
  (d) As customer encrypts his or her credit card with his or her private key it is confidential

5.43 When https is used
  (i) Credit card numbers are exposed to the merchant
  (ii) The merchant should encrypt credit card numbers before storing them in his or her hard disk
  (iii) Credit card numbers are not exposed to a merchant
  (iv) Credit card numbers are encrypted with customer's private key
    (a) (i), (ii)   (b) (i), (ii), (iii)
    (c) (ii), (iv)   (d) (iii), (iv)

5.44 The expansion of SSL is
  (a) Secure Server Layer
  (b) Secure Socket Layer
  (c) Simple Security Layer
  (d) Security at Server Level

5.45 SSL protocol layer is between
   (a) IP layer and TCP layer
   (b) Application layer, e.g., http and e-commerce application
   (c) MAC layer and TCP/IP protocol layer
   (d) TCP/IP layer and application layer

5.46 SSL provides
   (i) Encryption of transactions between a client and a server
   (ii) Server public key authentication
   (iii) A secure end to end channel between client and server
   (iv) Is implemented as a part of http
      (a) (i), (ii)   (b) (i), (v)
      (c) (ii), (iii)   (d) (i), (ii), (iv)

5.47 SSL uses for secure channel implementation
   (a) One-way hash function of transactions
   (b) A symmetric key encryption
   (c) RSA encrypted symmetric session key encryption
   (d) Clients private key encryption

5.48 https is an application which is
   (a) Independent of SSL
   (b) Insecure as it uses Internet
   (c) Available in clients and not servers
   (d) Requires SSL for encryption function

5.49 SSL
   (i) Does not require client public key certificate
   (ii) Requires public key certificate of server
   (iii) Allows use of any one of several encryption methods depending on the level of security required
   (iv) Requires client to have a public key certificate
      (a) (i), (ii)   (b) (i), (ii), (iii)
      (c) (ii), (iii), (iv)   (d) (ii), (iv)

CHAPTER 6

Payment Systems in E-Commerce

LEARNING GOALS

In this chapter, we will learn:
1. Various payment schemes and their requirements in e-commerce.
2. Secure Electronic Transaction (SET) protocol used for credit card payments.
3. Electronic funds transfer and e-cheques in B2B e-commerce.
4. Electronic cash payments.
5. Payment gateways and their role in e-commerce.
6. Payment method for information goods.

6.1 INTRODUCTION

In e-commerce just as in normal commerce it is necessary for persons/organizations availing of goods and services to pay for them. In all three forms of e-commerce, namely, B2C, B2B and C2C we need appropriate schemes of payment. In day-to-day commercial dealings there are primarily three modes of payment. The most common payment scheme between individual customers and merchants is cash, particularly for small value purchases. For larger value purchases usually a credit/debit card is the one accepted by most merchants. If you have a trusted relationship with a merchant he would normally accept cheque payment. In commercial dealings between businesses either cheque payments or instructions to banks to transfer amount due electronically (through Electronic Funds Transfer or Electronic Clearing System) is employed. The most common payment scheme in B2C e-commerce is by using credit cards. For small value transactions, less than Rs. 100, one would prefer cash payment. There are systems to mimic cash payment using the Internet which we will describe. There are several features of physical cash payment such as anonymity and zero transaction cost which are difficult to mimic in e-cash. Thus, what is available as e-cash is almost like cash but not identical to it.

In C2C systems, the transaction is through a broker. Thus, payment is handled by the broker either using credit/debit cards or by "cash on delivery" system. There are several requirements to be met by e-payment systems which will be described in the next section.

6.2 REQUIREMENTS OF E-PAYMENT SYSTEMS

There are several essential requirements and consequential requirements which should be met by e-payment systems. They are:

- Payment security which requires that any payment authorization is not tampered with by a hacker on the Internet.
- Privacy of transaction requires that third parties do not know for what goods and services one is paying. This also requires that the credit card number (transmitted over the Internet) is not stolen by an eavesdropper.
- The payment systems integrity should be assured. In other words, once an agreement is reached between a buyer and a seller neither can go back on their commitment.
- The customer and the merchant should be able to authenticate one another. In e-payment there is no physical contact between the two parties. There is no signed paper transaction. Thus, establishing mutual identities is essential.

Besides the above requirements electronic transactions must be designed to satisfy the following: (These are similar to properties of database transactions).

- *Indivisible*: Each payment transaction should be either whole or none. In other words, transactions should not be interrupted in the middle. If some malfunction occurs during a transaction, the whole transaction should be aborted and the state restored to the initial state.
- *Isolated*: Each transaction should be independent of others.
- *Agreed*: Both parties involved in the transaction should mutually agree on the terms and conditions.
- *Reversible*: If after conclusion of a transaction an error is found or if it is found that terms and conditions are not fully met, one should be able to reverse the payment and go to the initial state. For example, in credit card payment, if a customer is dissatisfied due to a valid reason the card company should credit the payment back to the customer and debit it from the merchant's account.

From the point of view of acceptability of a particular payment scheme for implementation the following requirements must be satisfied:

- *Standardized*: The system should be acceptable across computing platforms. In other words a universally accepted standard should be used to ensure inter-operability.

- *Economical*: Transaction cost of each transaction should be minimal (Ideally zero).
- *Scalable*: The system should be able to handle several transactions simultaneously. For example, if several customers login to a merchant's e-shop, the system should simultaneously service them while handling each customer as a separate entity.

The payment schemes which will be described in subsequent sections will be examined to satisfy ourselves that the above requirements are met.

6.3 CREDIT CARD PAYMENT

We described in the last chapter, the use of encryption to ensure security of transactions between a client and a server. As an example, we described one method of ensuring secrecy of credit card number when it is sent by a customer to a merchant using the Internet. There are many other steps involved before a credit card payment is complete. There are two systems which can be used. One is a simple system in which the credit card number is exposed to a merchant and the privacy of purchases made by a client is not ensured. The other system is more complicated in which the credit card number is not exposed to the merchant and the privacy of purchases is also ensured. The second method is more complicated and requires a customer to have a public key certificate which many casual customers may not have. This method, however, ensures that the credit card number is not exposed to a merchant and the privacy of transactions is ensured. We will describe both these methods in this section. Before describing these methods we will briefly review how credit card payments are made in physical shopping when a customer visits a shop to buy items. There are four parties involved in credit card transactions. They are:

1. A customer who owns a credit card.
2. A merchant who accepts credit cards (typically a merchant would accept credit cards of several companies such as Visa, Master card, etc.)
3. A bank which issues credit cards to customers, guarantees payment to merchants and collects bills from its customers.
4. An acquirer which is normally another bank which establishes an account with a merchant, validates card information presented by a merchant and approves sales based on a customer's credit status. The acquirer accepts cards of several credit card companies, transfers payment to merchant's account when it approves a sale and gets reimbursed by the issuing bank. The acquirer normally charges a commission of around 2% on each sale from the merchant. The customer's bank collects an annual fee from its customer and also a large monthly interest on outstanding overdue payments. Monthly interest may range from 1.5 to 3% on overdue bills.

From the above one can observe that the cost of administering the system is mainly borne by merchants.

Credit card transactions are carried out as follows (See also Figure 6.1):

1. A customer presents a credit card to a merchant after purchasing items from a store and agreeing to pay the billed amount.
2. The merchant swipes the card using a teleterminal which reads the data contained in the magnetic strip of the card and enters the transaction amount. The card data and amount are transmitted to the acquirer via a private communication line.

3. The acquirer's computer forwards the data to the bank which issued the card. The bank checks the validity of the card, credit available on the card and approves transaction provided the card and credit are OK.
4. The acquirer sends approval to merchant. The terminal at the merchant's premises prints a slip in duplicate approving the sale and the amount charged. The acquirer also credits the merchant's account with sale amount minus commission. The acquirer collects the amount from customer's bank.
5. The merchant requests the customer to sign the approval slip, compares the signature with that in the card and if OK delivers the goods.
6. The bank sends a monthly statement to the customer and collects the outstanding amount.

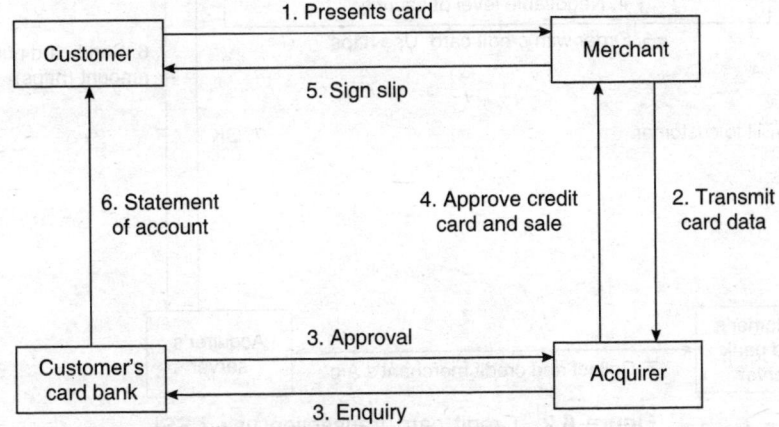

Figure 6.1 Manual credit card transaction.

Observe that there is physical proximity of a customer with a merchant and the transaction is validated after obtaining the signature of the customer on the payment slip. In e-commerce there is no physical contact between the merchant and the customer and it is impossible to verify the physical signature. It is also necessary for the merchant to verify that a customer is genuine and the customer to be assured that he or she is not dealing with a fake merchant. Thus, a customer would be reluctant to disclose his or her credit card number using the Internet as the merchant may be fake or the number may be stolen by eavesdroppers on the Internet. Furthermore, if the merchant is careless, a hacker may access his or her database and steal credit card numbers. There have been cases reported in the press of credit card numbers stolen from the merchant's database by disgruntled employees. An ideal protocol would be one in which the credit card number is not revealed to the merchant but only to the bank approving it. The approving bank need not know what a customer bought but only the amount of payment to be approved (to protect customer's privacy). Such a protocol has been standardized by major credit card companies. It is called *Secure Electronic Transaction* (SET) protocol. This protocol (as stated at the beginning of this section) requires every customer to have a public key certified by a certifying authority. This is not always possible as merchants have to accommodate casual customers. Another simpler protocol which uses Secure Socket Layer (SSL) will be described first. We will describe SET protocol later.

6.3.1 Credit Card Payment using Secure Socket Layer

The most common credit card payment prevalent in India nowadays is the one using the secure socket layer software implemented by most currently available browsers. The payment process using this method proceeds as follows (See also Fig. 6.2):

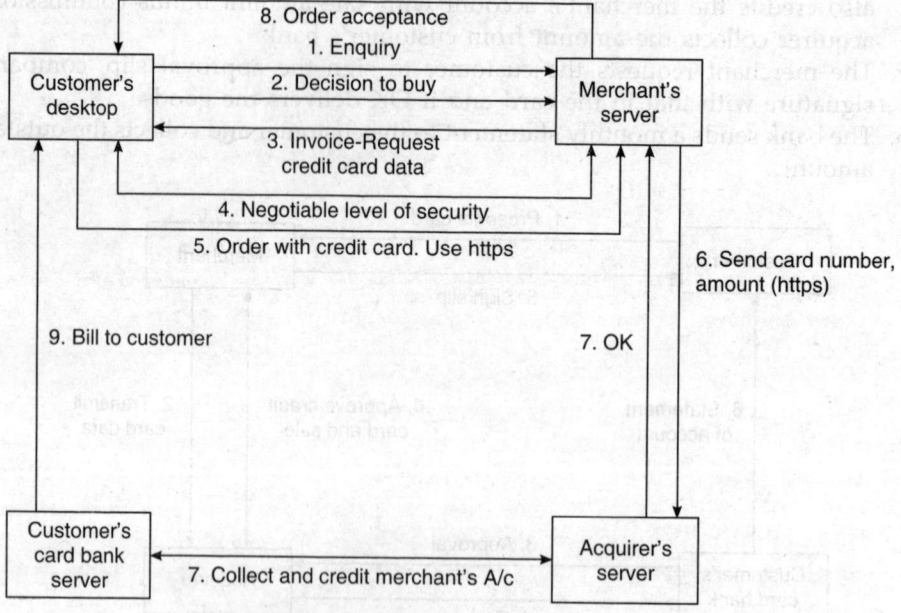

Figure 6.2 Credit card transaction using SSL.

Step 1: A client accesses the web site of a merchant by using its URL. A request to buy from a hypothetical company XX Bookstore will look like http://www.XXBookstore.co.in

Step 2: The client examines the catalogue of books, and selects the books he or she wants to buy. The selected items are placed in a virtual shopping cart assigned by the merchant's server.

Step 3: The book store's server examines the shopping cart assigned to the client and prepares an invoice including all taxes and shipping charges. This invoice is displayed in the client's browser along with a form requesting credit card information and shipping instructions.

Step 4: The server negotiates with the client's computer level of security. If the client's browser does not have SSL security the transaction cannot proceed as the credit card number will be exposed. A warning is displayed in the client's browser.

Step 5: Assuming client's browser is SSL compliant the client now types:

https://www.XXBookstore.co.in

As we saw in the last chapter https ensures that the data typed by the client is encrypted with a session key and sent to the web site XXBookstore.co.in. Thus, the credit card number and the shipping address are encrypted. As the session key is a random number and is used only for one session, eavesdroppers cannot use it even if they find it.

Step 6: The merchant decrypts the data. The credit card number, the amount to be paid by the customer, invoice details along with the merchant's public key certificate, is sent to the acquirer by the merchant encrypted using the acquirer's public key. The merchant also digitally signs the data. The details of invoice are sent to resolve any disputes between the customer and the merchant at a later date.

Step 7: The acquirer checks with the bank which issued a customer's card its validity, the customer's balance in this account and the merchant's digital signature. The acquirer will transfer the amount electronically to the merchant's bank and collects it from the customer's bank provided the customer's bank approves the transaction. It then sends authorization to the merchant to accept the sale. This authorization is encrypted with the merchant's public key.

Step 8: The merchant accepts the customer's order and sends an encrypted receipt to the customer. Later the merchant ships the items to the customer's shipping address.

Step 9: Finally, the customer's bank sends a monthly bill to the customer.

We see from the above description that:
1. The customer's credit card is exposed to the merchant. It is expected that the merchant will encrypt it and store it along with the invoice in his or her database.
2. The customer's purchase details are exposed to the acquirer. This is necessary to resolve disputes, if any, but the customer's purchase details will be exposed to the acquirer.
3. The major advantage of https protocol is that a customer need not have a public key. Sensitive data is encrypted by SSL and customer need not worry about eavesdroppers stealing his or her credit card number during Internet transaction.
4. The whole transaction is automated. The customer carries out the entire operation by clicks of the mouse button. Picking the encryption key, encryption, merchant authorization, etc., are all done by the https software (which has SSL underneath it).
5. If a customer uses http protocol instead of https, he or she will be warned about the website not being secure and in many systems the transaction will not be accepted.

This method is currently the most common one used in India for all types of e-commerce activities such as railway and airline ticket bookings, hotel reservations, e-shops, etc. We will describe in the next subsection a new protocol called Secure Electronic Transaction (SET) protocol.

6.3.2 Secure Electronic Transaction (SET) Protocol

Secure Electronic Transaction (SET) protocol has been standardized for credit card payments by major credit card companies such as VISA and MASTER CARD in USA. To use the SET protocol for credit card transactions, the following requirements must be satisfied:

1. As a public key encryption system such as RSA is used by both customers and merchants, both of them must have their own public–private key pairs.
2. Both customers and merchants must get their public key certified by a certifying authority. This is required to ensure to both parties that the transaction is genuine.
3. The customer must digitally sign the purchase order amount and credit card number. For digital signing a one way hash function such as MD5 should be used.

The main features of SET protocol are:
1. It ensures that a customer's credit card number is not disclosed to a merchant. It is disclosed only to the acquirer who authorizes payment.
2. Purchase invoice details are not disclosed to the acquirer. Only the credit card number and the total amount of purchase is sent to the acquirer.
3. Purchase invoice coupled with the credit card number is digitally signed by the customer so that an arbitrator can settle disputes, if any, on cost and purchase invoice.

The complete formal protocol definition has vast amount of detail (it is 262 pages long!). We will describe here only the essential aspects of the protocol. Readers interested in more detail may either see the web site www.ibm.com/redbook/SG244978 which has the entire standards document or read the book by Stallings given in the references at the end of the book [11].

6.3.3 Dual Signature Scheme

SET protocol depends on an innovation called dual signature whose main purpose is to give the merchant only the purchase order and amount without disclosing the credit card number, and give to the acquirer only the credit card number and the amount without disclosing the purchase details. This system will also ensure that the payment is as per the agreed purchase order amount. The essentials of the dual signature scheme is given in Figure 6.3.

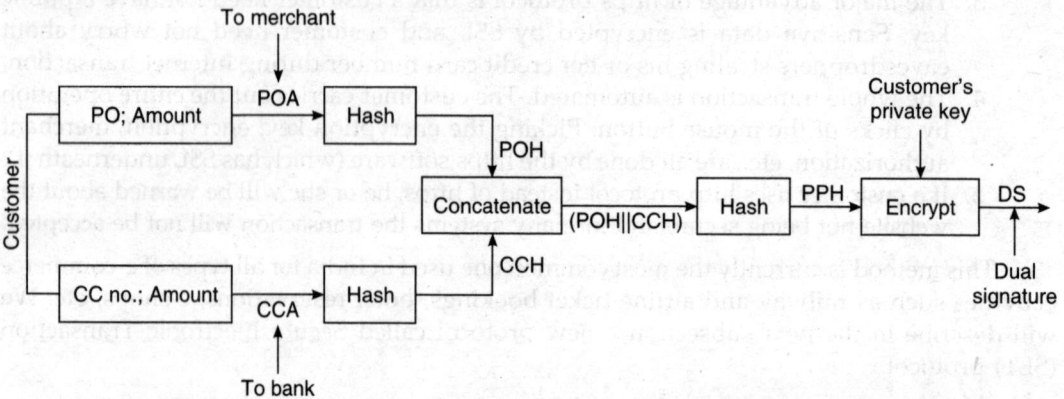

POA : Purchase Order; Amount; CCA : Credit Card No.; Amount; || : Concatenation operator;
POH : Hash of POA; CCH : Hash of CCA; PPH : Purchase Payment Digest

Figure 6.3 Essentials of dual signature scheme.

A customer's purchase information consists of a purchase order (PO) accompanied by a credit card number (CCN) and amount to be paid (AMT). This information is divided into two parts: (PO; AMT) = POA and (CCN; AMT) = CCA. The two parts are separately hashed using a standard hash algorithm such as MD5 explained in Chapter 5. Let Hash (POA) = POH and Hash (CCA) = CCH. POH and CCH are concatenated (i.e., stringed together) and hashed again giving PPH. Thus,

$$PPH = H(POH || CCH) \qquad (6.1)$$

where the symbol || is a concatenation symbol. PPH is encrypted with *customer's private key* K_{PR} giving

$$DS = K_{PR}(PPH) \quad (6.2)$$

DS is the customer's digitally signed copy of purchase order and credit card number. The purchase order and amount, namely, POA is separately encrypted by the customer using the *merchant's public key* and sent to the merchant. The merchant can decrypt it using his or her private key and obtain POA. The merchant also receives from the customer CCA encrypted with the *acquirer's public key*. Besides these CCH and DS are also sent separately to the merchant encrypted using the merchant's public key. The merchant can decrypt them with his or her private key and obtain CCH and DS. Remember that given CCH the merchant cannot find CCA as hashing is a one-way function. Thus, the credit card number is not available to the merchant. The merchant can compute:

$$H(H(POA) \;||\; CCH) = H(POH \;||\; CCH) = PPH \quad (6.3)$$

The DS received by the merchant can be decrypted by using the *customer's public key* to obtain

$$K_{PU}(DS) = K_{PU}(K_{PR}(PPH)) = PPH \quad (6.4)$$

(Remember that any quantity x encrypted with a private key can be decrypted with the corresponding public key to get back x).

K_{PU} is the certified public key of the customer which is sent to the merchant by the customer along with his purchase order. If H(POH || CCH) obtained in (6.3) equals K_{PU}(DS) obtained in (6.4) then the customer's signature is verified. If payment is authorized by the acquirer, the merchant can accept the order and ship it.

As the acquirer gets CCA encrypted with its public key it can decrypt it, encrypt it using the appropriate bank's public key and forward it to the bank. The bank can decrypt it and obtain CCA. The bank will also receive POH and DS. POA cannot be found from POH as POH is the hashed value of POA. The bank will thus not know the purchase details. It can, however, compute

$$H(POH \;||\; H(CCA)) = H(POH \;||\; CCH) = PPH \quad (6.5)$$

It will also receive K_{PU}(DS). If PPH obtained in (6.5) equals K_{PU}(DS), the digital signature of the customer is verified by the bank. If the credit available is sufficient, the bank can authorize the merchant (via the acquirer) to honour the purchase order.

Observe that the customer cannot repudiate his or her purchase order as it has been signed by him or her and deposited with the bank. The merchant also can not substitute a customer's purchase order with some other purchase order as the signature DS contains a unique hash of the customer's purchase order and this is available to the bank.

We will now summarize the procedure below (See also Fig. 6.4):

Step 1: The customer's PC and merchant's server exchange their certified public keys. The merchant also sends acquirer's public key to the customer. They also negotiate the type of public key encryption and the hashing function to be used.

Step 2: The customer fills the purchase order, amount payable and credit card number in his or her PC. Software in the PC strips it into two parts; purchase order and amount (POA) and the credit card number and amount (CCA).

POA is encrypted using merchant's public key and CCA with acquirer's public key. The PC also computes POH, CCH and the dual signature DS. These are also sent along with POA to the merchant. The merchant verifies signature and proceeds further if the signature is OK.

Step 3: The merchant forwards encrypted CCA, POH and DS to the acquirer.

Step 4: The acquirer forwards it to the customer's bank.

Step 5: The customer's bank checks the credit card number, credit available and the dual signature of the customer. The result of verification is sent to the acquirer.

Step 6: The acquirer in turn approves or rejects the transaction and informs the merchant. It credits (purchase amount–commission) in merchant's account.

Step 7: The merchant approves the order and sends to the customer the shipping details.

Step 8: At the end of the month the bank which issued customer's credit card sends a consolidated bill to the customer.

It should be remembered that all the operations are carried out by the appropriate software stored in the computers of each of the four participants, and as far as the customer is concerned the procedure is carried out by his or her PC by clicking the appropriate boxes using the mouse button.

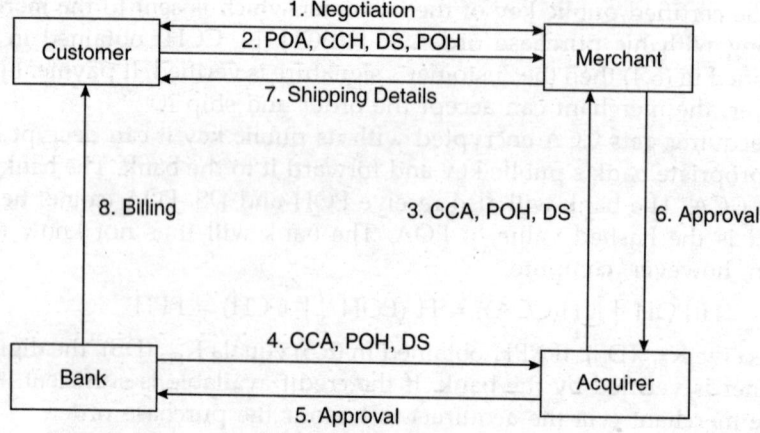

Figure 6.4 Credit card transaction using SET protocol.

6.4 ELECTRONIC FUNDS TRANSFER

Before describing cheque payment in e-commerce it is necessary to describe a scheme called Electronic Funds Transfer (EFT) which is an essential part of clearing cheques in a banking system. What do we mean by clearing a cheque? If A issues a cheque favouring B, it will be written on a cheque in A's bank (say X). B will deposit the cheque in his or her bank (say Y). Before B's account in bank Y can be credited, Y should check with A's bank X whether A has enough money in his or her account. If X approves then A's account in bank X will be debited and B's account in bank Y will be credited. This process of cheque clearance has been automated and is called Electronic Funds Transfer. In order to implement electronic funds transfer, the following requirements have to be met:

- An Automated Clearing House (ACH) should work as an intermediary to negotiate transfer of funds when cheques are used. In India, the Reserve Bank of India (RBI) acts as the ACH.
- All banks should use ACH and become its members. RBI is the controlling agency of all banks in India.
- There should be a secure electronic communication channel between each bank and the ACH. As the amounts involved are very large, normally these channels are private secure leased lines.

We will now examine two models of EFT. The first one we will call Automated Cheque Clearance and the second a variation called Electronic Clearing Service (ECS).

6.4.1 Automated Cheque Clearance

In the traditional model the following steps are followed (See Figure 6.5):

We will assume that A and B are the two parties involved and that A has an account with bank X and B with bank Y. ACH maintains balances kept by all its member banks.

1. A sends a cheque drawn on his bank X to B.
2. B deposits the cheque in his bank Y.
3. Y will send the cheque to ACH.
4. ACH has an automated system to sort cheques based on unique bank code. (In India it is the 9-digit bank code, e.g., 560002020 printed at the bottom of a cheque with magnetic ink). Based on X's code it sends a query to bank X whether the amount can be paid (A's physical cheque should be forwarded to X by ACH).
5. If the reply from bank X is yes, it debits X's account and credits Y's account with it.
6. It intimates bank Y that the cheque is cleared.
7. Bank Y credits B's account with the amount specified in the cheque and updates B's account.
8. Bank X debits A's account by the amount specified in the cheque.

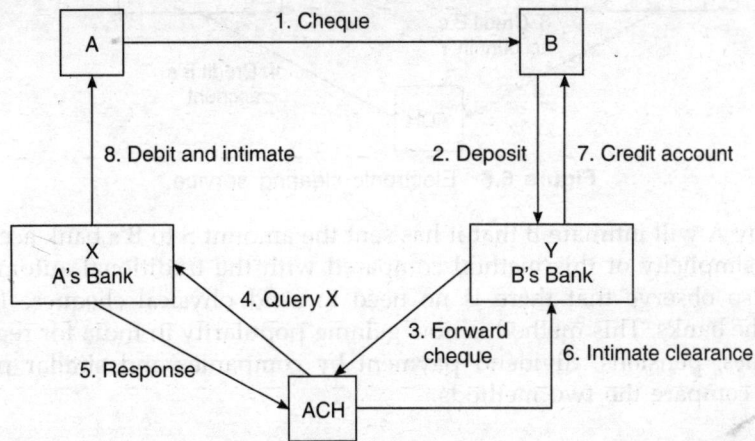

Figure 6.5 Automated cheque clearance.

In the traditional system, ACH primarily acts as a post office. The physical cheques are sent to the bank on whom it is issued for it to verify the cheque to authorize payment. The process normally takes a day or two. If physical cheques can be replaced by their electronic equivalent, the procedure will be faster and cheaper. We will examine in section 6.5 electronic cheques. The Electronic Cheque Clearance is an improvement as it eliminates the movement of physical cheques. Before that we will describe Electronic Clearing Service (ECS).

6.4.2 Electronic Clearing Service

In the Electronic Clearing Service (ECS), the following steps are followed (See also Figure 6.6). When A wants to send a specified amounts to B:

1. A requests B to send the unique 9-digit code of B's bank Y and B's account number in Y in which the money due has to be credited.
2. A sends an intimation to its bank X to debit its account by the specified amount and credit it to B's account with bank Y.
3. Bank X requests ACH to debit its account with ACH by S and credit it to Y's account. It also requests ACH to advise Y to credit B's account by the specified amount S.
4. ACH requests Y to credit B's account with the specified amount S.
5. Y intimates B that the amount from A has been credited.

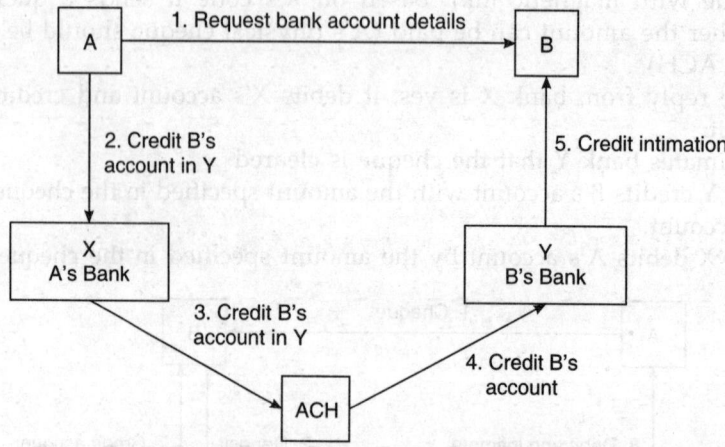

Figure 6.6 Electronic clearing service.

Normally A will intimate B that it has sent the amount S to B's bank account with Y. Observe the simplicity of this method compared with the traditional automated cheque clearance. Also observe that there is no need to send physical cheques. Thus, ECS is cheaper for the banks. This method is now gaining popularity in India for regular deposit of pay cheques, pensions, dividend payment by companies and similar payments. In Table 6.1 we compare the two methods.

Table 6.1 Traditional cheque clearance vs ECS

Traditional ACH cheque clearance	Electronic clearing service
Physical cheques transported	No physical cheques. Only e-advice
Delay of 1 or 2 days	No delay. Same day credit guaranteed
Cheque can be dishonoured if insufficient funds	No possibility of cheques being dishonoured
Transaction cost high due to physical handling of cheques	Low transaction cost as no physical transport of cheques
Payer can use fund during clearance delay	Payers account debited immediately

6.5 ELECTRONIC CHEQUE PAYMENT

In B2B e-commerce the purchaser and the vendor normally have mutual trust before negotiating a purchase. The amounts involved is also quite large. Thus, payments are normally made by cheque. Thus, we have to find an electronic equivalent of a physical cheque and payment scheme to mimic the manual scheme. We thus need banks and ACH to accept electronic cheques in addition to physical cheques. Assuming this condition is satisfied, we will describe a method.

Before electronic cheque payment system is implemented, the following conditions must be satisfied:

1. Both businesses must have certified public keys and must be able to digitally sign e-cheques. In India every registered company (Private or Public limited) must have a public key certificate and must be able to digitally sign documents. Income Tax department accepts only e-returns and electronic payments of tax due from all companies.
2. All banks and ACH must have public key certificates and must be able to digitally sign documents using a common agreed hash function.

There are five parties involved in the transaction.

1. Purchaser – P
2. Vendor – V
3. Purchaser's bank – PB
4. Vendor's bank – VB
5. Automated clearing house – ACH

One format for e-cheque is given in Table 6.2 which contains all the essential information.

Table 6.2 Format of e-cheque sent by purchaser (payer)

Unique id of e-cheque
Date of cheque
Payee' id and name
Amount to be paid
Payer's bank code
Payer's account number
Digital Signature of Payer

The steps followed in e-cheque payment are as follows (See Figure 6.7).

1. The purchaser (P) and Vendor (V) exchange their public key certificates and their respective bank details.
2. The Vendor (V) sends an invoice with payment request to the purchaser (P) encrypted with P's public key.
3. The purchaser decrypts the invoice and payment request. P checks validity of request and if OK writes an e-cheque (Table 6.2). The e-cheque along with invoice number and P's public key certificate are digitally signed and the digital signature is attached. This information is encrypted with V's public key and sent to the vendor (See Table 6.3). (This is called secure envelope).
4. The vendor verifies the signature and appends to this information an endorsement, his public key certificate and digitally signs all the above information. This is encrypted with the public key of the vendor's bank and sent to it (See Table 6.4).
5. The vendor's bank decrypts the information. It takes out the e-cheque, encrypts it with ACH's public key and sends it to ACH.
6. ACH decrypts it, forwards it to the purcahser's bank encrypting with PB's public key.
7. PB decrypts it and authorizes payment if all OK.
8. ACH debits PB's account and credits VB's account. It informs VB that the e-cheque is cleared.
9. VB credits vendor's account.
10. PB debits purchaser's account.

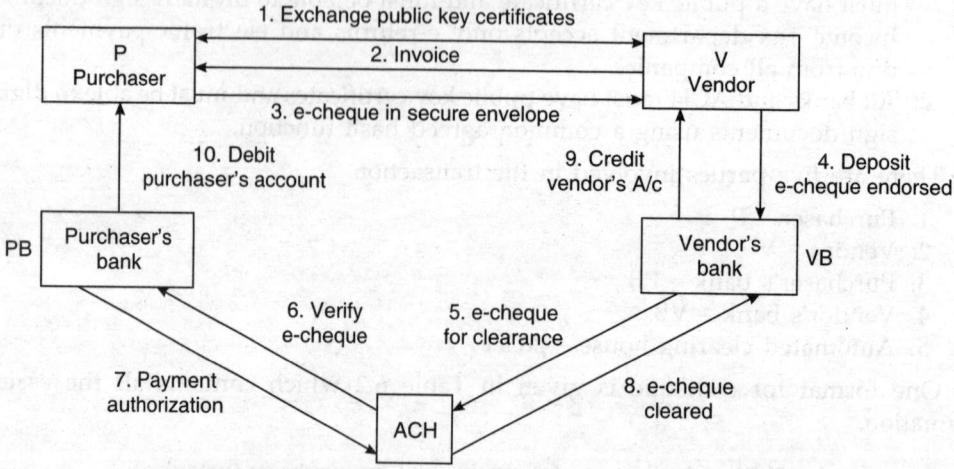

Figure 6.7 Electronic cheque clearance.

Table 6.3 Remittance in secure envelope

Invoice no.
P's public key certificate
e-cheque
Digital signature

Encrypt with V's public key

Table 6.4 Deposit slip to vendor's bank in secure envelope

P's public key certificate
V's public key certificate
e-cheque
V's endorsement with unique id
V's signature
Encrypt with VB's public key

6.5.1 Electronic Clearance of Pay Order

Electronic clearance of payment is available widely in India and is popular in B2B e-commerce. This works in a manner identical to the one described in Section 6.4.2. The parties involved are the same as in e-cheque payment. In this case also vendor, purchaser, vendor's bank, purchaser's bank, ACH must all have certified public keys and able to sign documents digitally using a commonly accepted hash function.

The steps followed in e-clearance of payment are as follows (See Figure 6.8).

1. The purchaser and vendor exchange their public key certificates.
2. The vendor sends invoice with amount payable to the purchaser.
3. The purchaser checks invoice and authorizes payment. (The purchaser will check electronically the amount payable against invoice. After authorizing payment, he or she will update account payable database). A pay order with details given in Table 6.5 will be sent to his or her bank PB encrypted with PB's public key.
4. The purchaser will intimate the vendor that ECS payment has been initiated.
5. PB will check accounts balance of the purchaser and if OK debit his or her account and send the information given in Table 6.6 to ACH encrypted with ACH's public key. It is assumed that ACH has certified public keys of all its member banks.

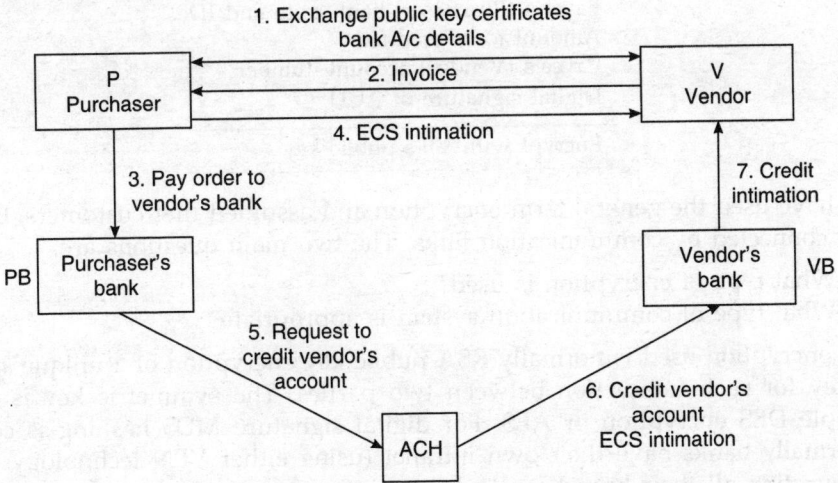

Figure 6.8 Electronic clearance of payment.

6. ACH will debit PB's account specified and credit VB's account. It will send information given in Table 6.7 to VB to credit the vendor's bank account.
7. VB will credit the vendor's account with payment received. It will intimate the vendor that ECS payment has been received and give payer's (purchaser's) ID.

Table 6.5 Pay order to purchaser's bank (contents of secure envelope)

Date on which to be paid
Unique ID of pay order
Purchaser's account number
Vendor's account number (Payee)
Amount to be paid
Digital signature of purchaser (Payer)
Encrypt with PB's public key

Table 6.6 Pay order by PB to ACH (contents of secure envelope)

Date of pay order
Unique ID of pay order
PB's bank code
VB's bank code and ID
Account number of payee (Vendor) in VB
Amount to be paid
Digital signature of PB
Encrypt with ACH's public key

Table 6.7 ACH's intimation to vendor's bank in secure envelope

Date of pay order
Unique ID of pay order
Payer's (Purchaser) bank code and ID
Amount to be credited
Payee's (Vendor) account number
Digital signature of ACH
Encrypt with VB's public key

We have used the general term encryption and assumed that customers, banks and ACH are connected by communication links. The two main questions are:

1. What type of encryption is used?
2. What type of communication system is appropriate?

The encryption used is normally RSA public key encryption of a unique symmetric session key for each transaction between two parties. The symmetric key is encryped using Triple DES encryption or AES. For digital signature MD5 hashing is commonly used. Normally banks have their own intranet (using either VPN technology or leased lines) connecting all their branches. Bank intranets are connected to one another using VPN which is a cheaper alternative to leased lines. ACH is usually connected to bank intranets using leased lines.

As amounts of money transferred between banks and ACH are tens of crores rupees every day it is advisable to use private leased line connection.

6.5.2 E-cheque Format

We have described the contents of e-cheque but not the exact format used for cheques. Unless a standard cheque format is used by all businesses, banks and ACH, it will be impossible for the programs running in each of the servers to automatically extract the fields of e-cheques, interpret them and process them. We have briefly discussed the need for an Electronic Data Interchange (EDI) standard for invoices, purchase orders and similar commonly used documents. EDIFACT (Electronic Data Interchange Format for Administration, Commerce and Transport) is standardized by the United Nations Economic Commission for Europe and has been adopted by many Government organizations in India. EDIFACT has a standard for financial instruments.

Another alternative is to use XML description of cheques. If all banks and ACH agree on a common XML definition, then e-cheques can be written using XML. XML is less expensive, easier to implement for interpretation by computers compared to EDI. In the current Internet and World Wide Web era, XML is a simpler alternative to EDIFACT.

6.6 ELECTRONIC CASH

In the previous sections, we described payment by credit cards and cheques. The cost of each transaction when a credit card or a cheque is used is high. It is estimated that each transaction costs at least five rupees. Thus, it is not appropriate for low value transactions, e.g., less than two hundred rupees. We normally use cash for such transactions. The major advantages of cash are:

- It is legal tender and its value is guaranteed in the short range. Inflation causes loss of value but is normally controlled by governments.
- It is universally accepted in exchange of goods and services without the intervention of third parties such as banks. A merchant who accepts cash can use it for buying goods and services without the intervention of the Reserve Bank of India.
- It is easy to carry around.
- It is anonymous. A merchant cannot say who gave a particular currency note.
- Privacy of transaction is ensured due to anonymity.
- Currency notes are breakable into smaller denominations. For example, if you buy an item for Rs. 44 and give Rs. 100 currency note to a merchant, he or she will return Rs. 56.

The major disadvantages of cash are:

- It is not safe. If cash is lost one would be lucky to get it back.
- Its bulk is proportional to the amount.

Governments do not normally print large denomination currency notes to prevent criminals from stealthily carrying large amounts of cash. The maximum denomination currency note printed in India is Rs. 1000.

The purpose of electronic cash (abbreviated to e-cash) is to mimic cash transactions with all its advantages and without its disadvantages. The major problems are:

- Who will issue e-cash? If it is not the Government, who will guarantee its acceptance? Who will guarantee its value?
- Will anonymity of e-cash be ensured? Should it be ensured? Law enforcement agencies in most countries discourage large e-cash transactions as anonymous cash is the favourite method used by smugglers, terrorists, corrupt officials and similar antisocial elements. Unlike cash which is bulky and can be detected when transported, large value e-cash can be transferred across national boundaries using the Internet and without detection if it is anonymous.
- If e-cash is issued by a bank will it be accepted universally? If a merchant sells goods for e-cash will he or she be able to use that e-cash to buy other goods? Will he or she be able to use it to return as "change" to another customer? The advantage of regular cash is this property of exchange.
- How will it be possible to detect forgery?
- How will it be possible to prevent double spending? In other words, a person who buys e-cash should not be able to spend the same more than once. When it is spent, it should lose its value or be taken out of circulation.
- How is the cost of handling e-cash recovered?
- How is the transaction cost made relatively low?

Currently e-cash is issued by some banks abroad. Most of the systems do not issue anonymous cash as law enforcement agencies discourage it. The amount of e-cash a person can buy is also limited to around Rs. 5000 as the primary purpose of e-cash in e-commerce is to pay for low value goods and services.

6.6.1 E-Cash Issue and Spending

We will describe in this subsection a generic method of issuing and using e-cash. Several systems have been designed. They are similar to the method, we describe in their essentials. The system is intended primarily for small cash transactions. The procedure is as follows (See also Figure 6.9).

1. A customer applies to a bank for e-cash depositing an amount in cash or by cheque. He or she requests the bank to issue e-coins in various denominations such as Rs. 100, Rs. 50, Rs. 10, Rs. 5, Rs. 2, Re. 1 for the amount deposited. The bank will limit the total amount to around Rs. 5000. (It is assumed that a merchant with whom a customer wants to transact business has an e-coin account with this bank).
2. The bank has a currency server which issues e-coins. It assigns to each coin a random number as ID, notes its denomination and digitally signs them with its private key. E-coins also have an expiry date. The bank maintains a database of issued coins with a unique customer ID. This database is also sent to the customer's PC using SSL connection and stored in an e-coin database called e-purse. The essentials of the issued coin database is shown in Table 6.8.
3. The customer issues an e-coin from his or her e-purse to a merchant for items ordered. The e-coin is sent to merchant using https protocol to ensure security. The customer's PC marks the coin as spent in its e-purse.
4. The merchant sends this e-coin to the bank's currency server using SSL connection. The currency server verifies its signature in the e-coin and whether the e-coin is already spent or expired. If it is not spent or expired, it accepts the e-coin and changes the spent field from No to Yes in the issued e-coin database.

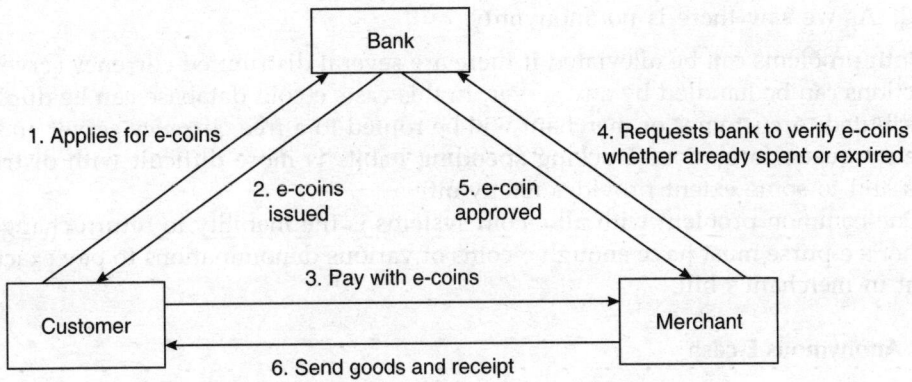

Figure 6.9 Electronic cash issue and payment system.

Table 6.8 Issues coins database

Customer ID:

Coin ID	Amount	Digital Signature	Expiry Date	Spent
2894	100	$XYbEy	08/10	No
2948	50	ZYXAB$@	08/10	No
4827	20	AXBYCZDY	08/10	No
⋮	⋮	⋮	⋮	⋮

5. The result of its verification, namely, whether e-coin is accepted or rejected is sent to the merchant. It also credits the merchant's account maintained by it by the approved amount (less commission).
6. The merchant ships the item ordered to the customer and sends a digitally signed receipt.

Observe that this scheme does not preserve the anonymity of the customer as the customer has an account with the bank and applies for issue of e-coins revealing his or her identity. The bank can also monitor the spending habits of a customer. The merchant cannot exchange the e-coins with another merchant as the bank has to verify the validity of the e-coin. The process of validation invalidates the e-coin to prevent double spending. E-cash can only be given to those merchants who have an account with the bank which issued e-coins to a customer. Thus, it is not like cash which is universally accepted. The bank uses its signature to detect forgeries. The cost of issuing and managing e-cash is recovered from merchants.

A more complicated version of e-cash requires the customer to digitally sign the e-cash they issue. This can be decrypted by the merchant using the customer's public key. The merchant also sends the e-coin digitally signed to the bank. The idea of both the customer and merchant digitally signing e-coins is to ensure that neither can repudiate their transactions at a later time.

There are two problems with the above scheme:

1. If there is a single currency server and there are a large number of customers and merchants, then the system can encounter a large delay if many clients simultaneously log on to the server.

2. As we saw there is no anonymity.

Both problems can be alleviated if there are several distributed currency servers and transactions can be handled by any server. In this case, e-coin database can be duplicated or distributed. A customer or merchant will be routed to a free currency server and there will be no queue formation. Tracking spending habits is more difficult with distributed servers and to some extent provides anonymity.

One common problem with all e-coin systems is the inability to return change. The customer's e-purse must have enough e-coins of various denominations to pay exactly the amount in merchant's bill.

6.6.2 Anonymous E-cash

In the method we described so far an issuing bank can link a customer with a merchant and will thus know the spending habits of a customer. There is a protocol called Chaum's blinding protocol which can in theory allow a customer to buy e-coins and blind them (i.e., make them anonymous before spending). The steps in the procedure are: (We assume all transactions are via the Internet and all communications use SSL connections.)

1. A customer applies for a specified amount of e-cash to an e-cash issuing bank and requests the bank to debit this amount from his or her account.
2. The customer creates e-coins of required denominations for the total amount identifying each e-coin with a unique random serial number which has at least 50 digits.
3. The customer picks a random number and encrypts it with the bank's public key and multiplies this encrypted number with the random serial number of the e-coin and sends it to the bank's currency server. Thus, the original serial numbers are not known to the bank. The e-coin is thus "blinded".
4. The bank's currency server signs the denomination of the e-coin and encrypts the blinded number with its private key. This constitutes its signature. The entire data is encrypted with the customer's public key and sent to him or her.
5. The customer decrypts it, divides the "blinded" serial number by the random number he or she used for blinding and gets the original serial number of the e-coin, the amount and bank's signature of the original serial number and amount. This now constitutes the customer's e-coin database or e-purse. Each coin has the form: (Serial number of e-coin, amount signed by bank, serial number signed by bank). He issues e-coins to a merchant when he or she buys goods.
6. The merchant sends the e-coins to the bank's currency server which decrypts the signature and verifies that the e-coin is authentic. From the serial number it cannot make out to whom it had issued the e-coin. It can store the coin with its serial number in a spent coin database. If the same coin is presented again, it can suspect double spending. It cannot, however, easily link the serial number with a specific customer. However, as random serial numbers are generated independently by each customer and they are more than 50 digits long, the probability of two customers generating the same serial number is low. In case duplicate is presented to the currency server Chaum (the inventor of the scheme) has developed a protocol in which the bank's currency server can challenge the

customer to prove that the e-coin is legitimate. If the customer is not satisfactorily able to respond, the bank knows the identity of the customer and can reject the e-coin and penalize the customer.

It is seen from the above description that it is time-consuming to trace double spending to a specific customer. Thus, banks are reluctant to issue anonymous e-cash. In online transactions time-delay will make the server slow.

The basic mathematics of Chaum's blinding protocol is explained below for those who are interested:

Let (n, e) be bank's public key and (n, d) be bank's private key.

1. Customer sends to bank

$$[s\ r^e\ (\text{mod}\ n), a]$$

where s is the serial number of the e-coin and a its denomination, r a random number called the blinding factor.

2. Bank signs this with its private key giving

$$[(s\ r^e)^d\ (\text{mod}\ n), a^d\ (\text{mod}\ n)]$$

and sends this to the customer.

However,

$$(s\ r^e)^d\ (\text{mod}\ n) = (s^d\ r^{ed})\ (\text{mod}\ n)$$

As $r^{ed}\ (\text{mod}\ n) = r$, the signed serial number is $(s^d\ r\ (\text{mod}\ n))$.

The customer divides this by r which is known to him or her getting

$$s^d\ (\text{mod}\ n)$$

which is the original serial number signed by the bank. Signed e-coin stored by the customer is:

$$[s, a, s^d\ (\text{mod}\ n), a^d\ (\text{mod}\ n)]$$

which is [serial number of coin, amount, serial number signed by bank, amount signed by bank]

This is presented to the merchant. The merchant as well as the bank can find the authenticity of this coin as both know the bank's public key.

Two points need to be observed:

1. The bank signs $(s\ r^e)$ from which it cannot find s, the serial number of the e-coin. However, from this the customer gets the serial number signed by the bank as he or she knows r.
2. The customer can use a different random number for each e-coin making it more difficult for the bank to guess the random blinding factor.

Apart from the complexity of the algorithms used to make anonymous e-coins work, law enforcement authorities in most countries discourage use of anonymous e-cash as it is often used for illegal purpose such as evading tax.

6.6.3 Smart Card-Based Cash Payment

Smart cards are plastic cards with embedded memory and a small special purpose processor. Unlike credit cards which store data on a magnetic strip which cannot be altered, smart cards can store data which can be altered. Data can be read or written in the card by a device called a smart card reader/writer which is normally connected to a PC. There are several types of smart cards such as stored value cards used for payment, SIM cards (used in mobile phones) to store prepaid amount, phone numbers, ringtones, etc., and access control cards to enter buildings. There are two types of cards—contact cards and contactless cards. Contact cards have gold plated terminals which make electrical contacts with connectors in the smart card reader. SIM cards used in mobile phones are of this type. Contactless cards have an embedded antenna. When the card is taken near a reader it transfers data by wireless coupling (more accurately electromagnetic coupling) to the reader.

Another classification of cards are those which have only memory and those which have memory and a processor. The memory in the memory card is usually divided into a read only part and a read/write part. The read only part is used for data such as passwords or unique identification number and the read/write part for variable data such as value stored.

One method of cash payment using the Internet in e-commerce is to use smart cards which store money value. Such cards are issued and replenished by some banks on payment of appropriate amount. A user equipped with a smart card reader attached to a PC can read the contents of the card (See Figure 6.10) and enter on the keyboard the cash to be paid. The amount in the card is debited as soon as the amount is entered. The card details and amount are sent encrypted (using SSL) to a merchant. Merchants transfer this data (after encrypting it) to the issuing bank. The issuing bank checks the validity of the card, (expiry date, bank's signature, etc.) balance amount in it and conveys approval or rejection to the merchant and credits merchant's account if approved. The merchant ships the merchandise and sends shipping details and receipt to the customer. The major difference between this and e-coins is that the card gets debited as soon as the money is spent and the bank need not check double spending. In a smart card with embedded processors, the public key and private keys are stored. Encryption can be done by the processor in the card to ensure security. The main disadvantage is the need for a customer to have a smart card read/write device on his or her PC. The processing cost of the transaction is a little

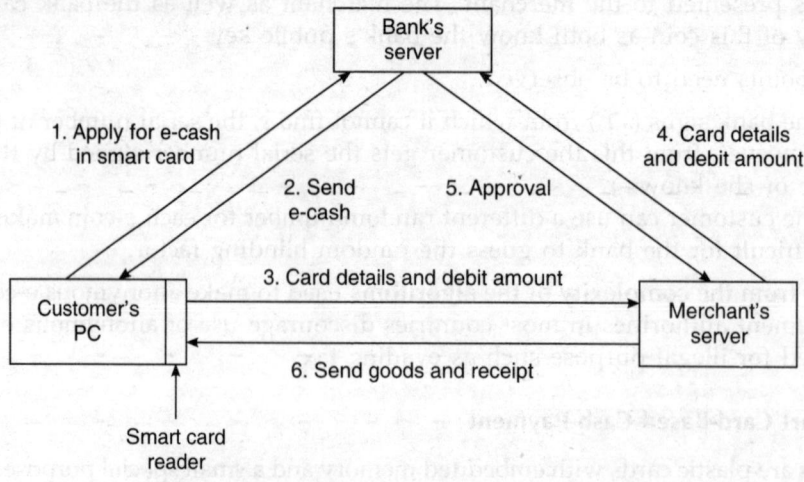

Figure 6.10 Smart card-based payment system.

lower than that of a credit card. Remember that the customer pays the bank in advance before spending the money. Thus, the bank earns interest on the unspent amount and to some extent it recovers the transaction cost. Further, the bank normally charges the merchant a transaction cost (which maybe lower than the charge for a credit card transaction). A customer also finds use of smart card advantageous as it can be replenished by the bank remotely. The card is online only when used and thus safe from hackers and the stored value is limited. The card can be "locked" with a personal identification number (PIN) so that even when it is lost it cannot be used. The disadvantage is, it is not anonymous as the bank while checking the validity of the smart card can link the payment to a customer. Anonymous e-cash smart card has been proposed using Chaum's blinding algorithm.

6.7 PAYMENT GATEWAYS

Many Small and Medium Businesses (SMBs) do not have programmers to design their web server to transact credit card, cheque or ECS transactions. Thus, several companies have designed what are known as *payment gateway solutions* to assist such SMBs to transact Internet-based payments. Primarily the payment gateway is a server which is placed between the merchant's web server and the acquiring bank's server. (See Figure 6.2). When an order is placed with a merchant it is forwarded to the payment gateway server by the merchant's server. The payment gateway server performs all the tasks such as integrating merchant's shopping cart with billing system, order approval, calculating appropriate taxes, forwarding shipping details, etc. It forwards the credit card information to the acquirer and gets backs the result of credit card purchase authorization. It also collects credit card payments and remits them to the merchant's bank account periodically. It checks customer's credentials, address verification and fraud prevention as an added service.

In the case of ECS payment in B2B e-commerce the payment gateway has arrangements with several banks to enable transfer of funds from one member bank's account to another member bank's account. In other words it functions as a limited electronic funds transfer agent. It often provides Internet banking solutions to member banks and their customers. Gateway operators collect a percentage of transaction amount as fee.

In India there are several payment gateway operators. Some of the leading ones are CC Avenue, ICICI Pay Seal and CC Now.

6.7.1 Pay Pal

Pay Pal is a service somewhat like an Internet based bank. The PayPal.com site operates accounts for members in which members can deposit money and transfer money to other members of Pay Pal. Other services provided by Pay Pal are as follows:

1. Pay for items bought in e-bay auctions (important in C2C e-commerce).
2. Transfer money to other Pay Pal accounts located in several countries where there are no exchange controls.
3. Buy things from web sites of merchants who maintain Pay Pal account.
4. It has also a mobile phone-based system offering similar services.

The primary use of Pay Pal is in C2C e-commerce among its members. It can also be used in B2C e-commerce if merchants maintain Pay Pal accounts. Pay Pal provides an alternative to those who do not have a credit card.

6.8 MICRO-PAYMENTS FOR INFORMATION GOODS

Micro-payments are small payments of a few rupees or dollars. Information goods are materials such as e-books, e-papers, audio (e.g., music) files, video entertainment clips and software files. A system to collect micro-payments should have low transaction cost. A system operating in USA has been implemented by a company named Net Bill in cooperation with Mellon Bank (a USA bank). The main features of this payment system are:

1. Customer deposits in advance an amount in Net Bill's bank account maintained by Net Bill server.
2. This deposit is debited only when the information goods is delivered to the customer.
3. The vendor is guaranteed payment when information goods are delivered as per customer's order.

The payment scheme has nine steps given below (See also Figure 6.11):

1. A customer requests a quote from the merchant for the item he or she requires.
2. The vendor responds with a quote for the item.
3. The customer informs the vendor of acceptance (If it is not accepted, no further transactions are needed).
4. The vendor encrypts information goods using a secret key and sends it to the customer. The customer cannot decrypt the information and use it until he or she gets the decryption key.
5. The vendor sends the key used for encrypting the information to the Net Bill server.
6. The customer sends a debit note along with digest of the information received to the Net Bill server using SSL connection. The digest is sent to settle any disputes that may arise between the customer and the vendor. The Net Bill server will keep the digest for a limited period of time (around two weeks). (It is assumed that the customer has enough funds to cover the debit. If not, Net Bill's server informs the customer and requests the requisite money to be credited to his or her account.)

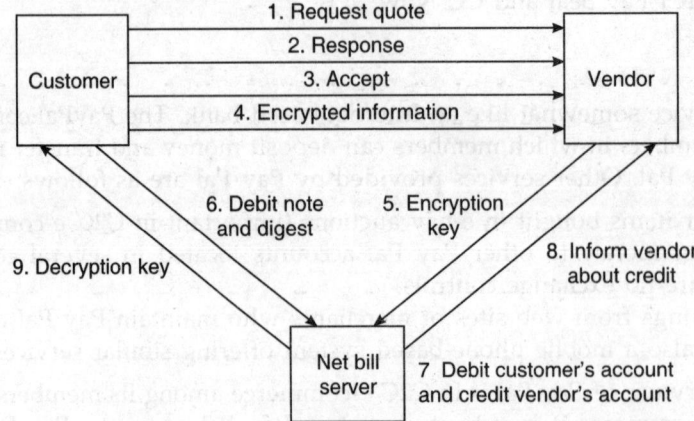

Figure 6.11 Net Bill system for information goods.

7. The Net Bill server credits the vendor's account by the amount debited from the customer less a transaction fee.
8. Net Bill informs the vendor of credit.
9. The Net Bill server sends the customer the decryption key to enable him or her to use the information.

This method has a lower transaction cost compared to credit card payment as the customer is debited from money already deposited with Net Bill. Net Bill protects both the customer and the vendor from frauds by keeping the digest of information delivered and releasing the decryption key only when the payment has been credited to the vendor's account.

6.9 CONCLUSIONS

In this chapter, we saw how payments are made by customers to vendors in e-commerce. Three major methods of payments are by credit cards, cheques and by cash. Of these currently credit card-based payment is most popular as it is well established and widely implemented in all countries. In B2B e-commerce large amounts are involved and credit cards are not appropriate. Electronic Clearing Service is the simplest one to implement. It can be implemented by cooperating banks using a middleman. If both parties have accounts in different branches of the same bank (such as State Bank of India), then most banks have an intranet connecting all their major branches and direct credit from one account to another is possible. Most countries have a Central bank which acts as an intermediary accepting e-cheques from various banks, clearing them and depositing them in specified accounts. Small payments are done using e-cash by debiting cash accounts kept by some banks or by using smart cards which store prepaid amount which are debited. In Table 6.9, we compare various payment options and appropriateness in different situations.

Table 6.9 Comparison of e-payment methods

Property	Credit Card	e-cheque	e-cash
When payment made	Later at billing time	When cheque is realized	Prepaid
Safety	Credit card company bears risk partially	Can stop payment if fraud suspected	Not good
Anonymity of purchase	SET protocol ensures anonymity	No anonymity	Most systems provide partial anonymity
Transaction cost	High	Medium	Low
Transaction value	Upto credit limit in card	Upto balance in account	Prepaid limit. Cannot give change
Popularity	Most popular in B2C e-commerce	Used in B2B e-commerce only	Not implemented worldwide. Pay Pal popular in C2C e-commerce
Security	SSL or SET	SSL or VPN	SSL
Integrity	Good	Good	Good
Non-repudiation	Good	Fair	Good

SUMMARY

1. In e-commerce there are several payment methods which include credit card payment, cheque payment, cash payment and smart card (or debit card) payments.
2. Credit card and cash payments are normally used in B2C e-commerce. Cheque payments are used in B2B e-commerce. An intermediary known as a payment gateway is used in B2C e-commerce by small and medium businesses.
3. All payment systems must ensure security, privacy, integrity of transaction and non-repudiation of payment received.
4. Each payment transaction should be indivisible (all or none), isolated from one another, mutually agreed upon and reversible.
5. Payment schemes should be standardized so that they are accepted in any computing platform, must be economical and must be scalable so that the system handles several payments simultaneously.
6. In credit card payments four parties are involved. They are: 1. A customer who uses a credit card for payment. 2. A merchant who accepts it. 3. An acquirer who accepts credit cards of several card companies, approves sale by the merchant and guarantees payment to the merchant. 4. The customer's credit card issuing bank which bills the customer and collects the amount due.
7. In e-commerce all transactions use the Internet. Thus, ensuring security is of paramount importance. There are two protocols for credit card payments. One of them is simple and most commonly used. This protocol meets all the requirements of e-payment systems except privacy. The other has been standardized by major credit card companies. It is called SET protocol. It is complicated but ensures privacy. It is not yet widely used in India.
8. The SET protocol uses a dual signature scheme in which the merchant is not disclosed the credit card number but only the purchase order and amount. The acquirer, on the other hand, knows only the credit card number and the amount and not purchase details.
9. Cheque payment requires a scheme called Electronic Funds Transfer (EFT). In this scheme e-cheques are sent to an intermediary who communicates with the payer's bank to ensure that the payer has enough funds and transfers the amount to the payee's bank account.
10. There is another system called Electronic Clearing Service (ECS). In this system the payer requests his bank to transfer the amount in the cheque to the payee's bank account. This is done using the services of an intermediary to which the buyer's bank sends instructions to transfer a specified amount to payee's bank account.
11. In B2B cheque payments, digitally signed e-cheque is issued by a payer. Through an intermediary (using either EFT or ECS) the amount is transferred to the payee. As all communications use the Internet, public key encryption and digital signatures are used in all the transactions.
12. A standard format for e-cheques is essential for facilitating automated transaction processing. A format called Electronic Data Interchange Format has been

standardized by EDIFACT. Another alternative is to use XML to define the format which can be interpreted by the intermediary.

13. Electronic cash is used for small payments to merchants and is cheaper than credit card transactions. It is difficult to emulate all characteristics of real cash in e-cash. However, a system is available in which a bank issues e-coins to a customer signed by the bank which is accepted by merchants who also have accounts with the same bank.

14. There are two systems. One of them does not ensure anonymity of spending habits and the other ensures anonymity.

15. In the non-anonymous system, an e-cash bank issues e-coins of specified denominations identified by unique numerical id, signs it and sends it to the customer. The customer pays the merchant using e-coins. The merchant forwards the e-coins to the bank's server which checks the signature and whether it is already spent. If the e-coin is OK, it informs the merchant who accepts the e-coins and executes the order.

16. In anonymous e-cash, the customer creates e-coins of various denominations, assigns a random 50-digit serial number to each e-coin, blinds it with another random number and forwards it to the bank's server. The bank signs the blinded e-coins and returns them to the customer. The customer "unblinds" the e-coin and gets the bank's signed serial number and denomination of e-coin. The rest of the steps are similar to that used in non-anonymous e-cash. The only complication is the difficulty in checking double spending of the e-coin. Details of blinding method are given in the text.

17. Payment gateway is a server with software provided to merchants by some companies. The gateway handles all credit card and cheque transactions which come to the merchant's web server, processes them and forwards them to the acquirer's server or electronic funds transfer server of banks. Payment gateways also provide services to banks to allow them to provide e-banking facilities to their customers.

18. Pay Pal is a popular service which acts as an e-bank maintaining accounts of several customers and merchants. E-payment among their customers is handled securely by this service. It is particularly useful in C2C e-commerce among its members.

19. A system has been implemented by a company called Net Bill (in USA) for handling small payments for information goods such as text, audio or video. A vendor delivers encrypted information to a customer and sends the decryption key as soon as payment has been credited to its account by Net Bill.

EXERCISES

6.1 What are the requirements of payment systems in e-commerce?
6.2 What is the difference between payment security and payment integrity?
6.3 What do you understand by privacy of payment in e-commerce?
6.4 Why is authentication required in e-commerce?

6.5 What properties are essential in e-payment transaction?

6.6 What do you understand by reversibility of transaction? Why is it required?

6.7 What requirements must be met for general acceptability of an e-payment system by most entities participating in e-commerce?

6.8 What do you understand by scalability of e-payment system?

6.9 Enumerate the entities involved in a credit card payment system and their individual roles.

6.10 How are credit card transactions carried out in normal everyday shopping? What are the special problems which are encountered when credit card transaction is to be carried out in e-commerce?

6.11 Explain the simple credit card transaction carried out in e-commerce using only Secure Socket Layer. What are its drawbacks?

6.12 In the method explained by you as answer to Exercise 6.11 is the credit card number exposed to the merchant? Is there any problem with transaction privacy?

6.13 What is SET protocol in credit card transactions? What are its unique features?

6.14 What is dual signature? What is the purpose of dual signature in SET protocol?

6.15 Explain with a block diagram how credit card transactions using SET protocol works.

6.16 How does SET protocol ensure security of credit card details?

6.17 How is privacy of transaction ensured when SET protocol is used?

6.18 What is electronic funds transfer? How are cheques cleared in this type of transfer?

6.19 What is an automated clearing house? What is its role in e-cheque payment system?

6.20 In day-to-day physical cheque transaction what is the role of Reserve Bank of Inda? Explain how a cheque issued by A to B gets credited to B's bank account.

6.21 How does an e-cheque issued by business A to B get credited to B's account? Explain all the steps.

6.22 What is the difference between ECS and EFT? From the point of view of a business issuing a cheque which is advantageous? What is the advantage if any from the point of view of a business receiving ECS payment?

6.23 How is security of e-cheque payment ensured?

6.24 What is an electronic pay order? Give its format. How is a pay order processed?

6.25 Enumerate the steps in carrying out ECS of pay order using the Internet. How is security of transactions ensured?

6.26 What type of encryption and communication are used in electronic clearance of pay orders?

6.27 Is there is a standard used for e-cheque format? Find out from the web if there is any internationally accepted format.

6.28 What are the advantages of cash in normal commerce? If we want to replace cash by e-cash for e-commerce what are the ideal characteristics of real cash should it meet?

6.29 Describe an e-cash system which is not anonymous. What characteristics of real cash does it meet?

6.30 Give the algorithm used to blind e-coins in anonymous e-cash.
6.31 What is a smart card? In what way is it different from credit card?
6.32 How does a smart card store cash value? How is it recharged when cash stored in it is exhausted?
6.33 Distinguish between contact smart cards and contactless smart cards.
6.34 Is it essential for all smart cards to have an embedded microprocessor? When are embedded microprocessors useful?
6.35 Are smart card payments anonymous? If not, explain why they are not anonymous.
6.36 What are payment gateways? How does a payment gateway assist small and medium businesses to participate in e-commerce.
6.37 Enumerate the various services provided by payment gateways.
6.38 What is Pay Pal? What are the services provided by Pay Pal? In which type of e-commerce is it found most useful?
6.39 What are the requirements of an e-payment system for information goods? How does the Net Bill system meet these requirements?
6.40 What is the purpose of sending only encrypted information to a customer in the Net Bill system? How does the customer protect his or her interest if there is a need for dispute settlement?

OBJECTIVE QUESTIONS

Each question has four possible answers. Pick the most appropriate answer.

6.1 The most popular e-payment system in B2C e-commerce is
   (a) Credit card payments
   (b) E-cheque payments
   (c) Smart card based payments
   (d) E-cash payments

6.2 Privacy in e-payments imply that the
   (a) Merchant does not know what a customer has purchased
   (b) Other customers do not know what a customer has bought
   (c) An entity other than the merchant does not know what the customer bought
   (d) A merchant does not know what a customer bought from another merchant

6.3 A payment system must meet the following requirements:
   (i) Payment security
   (ii) Payment privacy
   (iii) Mutual authorization of the two parties involved in the transaction
   (iv) Non-repudiation
      (a) (i), (ii)
      (b) (i), (ii), (iii)
      (c) (i), (ii), (iv)
      (d) (i), (ii), (iii), (iv)

6.4 By non-repudiation, we mean when an agreement is reached between a customer and merchant on a purchase
   (a) Both should honour it
   (b) The merchant should honour it
   (c) The customer should honour it
   (d) Neither is compelled to honour it

6.5 By payment system integrity, we mean
   (a) Integrity of transaction
   (b) Non-repudiation of agreed transaction
   (c) Privacy of transaction
   (d) Correctness of transaction

6.6 Individual payment transactions must have the following properties:
   (i) Reversible
   (ii) Isolated
   (iii) Agreed
   (iv) Indivisible
       (a) (i), (ii)   (b) (i), (ii), (iii)
       (c) (i), (iii), (iv)   (d) (i), (ii), (iii), (iv)

6.7 By transaction reversibility, we mean
   (a) That the transaction is reversible
   (b) Once an agreement is made it should not change
   (c) If an error occurs during a transaction one should be able to return to the initial state
   (d) All transactions should be stored and be reversible

6.8 An e-payment scheme becomes generally acceptable if it is
   (i) Economical
   (ii) Scalable
   (iii) Standardized
   (iv) Good
       (a) (i), (ii)
       (b) (i), (ii), (iii)
       (c) (i), (iii), (iv)
       (d) (i), (ii), (iii), (iv)

6.9 An e-payment scheme is scalable if
   (a) It can be expanded
   (b) It can be generalized
   (c) It is able to handle several transactions simultaneously and independently
   (d) No bottleneck develops in the system

6.10 In credit card transactions an acquirer
   (a) Interacts with the customer
   (b) Accepts credit card details and amount from merchants and validates and approves transaction
   (c) Forwards the credit card to a bank
   (d) Debits customer's account

6.11 In credit card transactions an acquirer
   (a) Accepts credit cards of several credit card companies from merchants
   (b) Accepts credit cards of a selected set of credit card companies from its members
   (c) Accepts only debit cards as it debits customers account by purchase amount
   (d) Accepts transactions from member banks

6.12 In credit card transactions using Secure Socket Layer Internet connection, the credit card number is
(a) Not revealed to a merchant
(b) Revealed to the merchant
(c) Not revealed to the acquirer
(d) Revealed to the merchant and not the acquirer

6.13 In credit card transactions using SSL Internet connection, the purchase details are
(a) Not revealed to the acquirer
(b) Revealed to the acquirer and merchant
(c) Not revealed to the merchant
(d) Revealed only to the merchant

6.14 In credit card transactions using SSL Internet connection, a customer
(a) Need not have a certified public key
(b) Requires a certified public key, private key pair
(c) Requires a private key to sign the purchase order
(d) Needs a public key

6.15 When SET protocol is used in credit card transactions in e-commerce, credit card details are revealed to
(a) The merchant only
(b) Both the merchant and acquirer
(c) Only to the acquirer
(d) Neither merchant nor the acquirer

6.16 When SET protocol is used, the details of purchase are revealed to
(a) The merchant only
(b) Both the merchant and the acquirer
(c) Only to the acquirer
(d) Neither the merchant nor the acquirer

6.17 The Secure Electronic Transaction protocol is used for
(a) Credit card payments
(b) Cheque payments
(c) Electronic cash payments
(d) Payment of small amounts for Internet services

6.18 In SET protocol a customer encrypts the credit card number using
(a) His or her private key
(b) The acquirer's public key
(c) Bank's private key
(d) Merchant's public key

6.19 In SET protocol a customer sends a purchase order
(a) Encrypted with his or her public key
(b) In plain text form
(c) Encrypted using merchant's public key
(d) Using digital signature system

6.20 One of the problems with using SET protocol is
(a) Merchant's risk is high as he or she accepts encrypted credit card
(b) Credit card issuing bank should check digital signature

(c) Credit card issuing bank has to keep a database of the public keys of all its customers
(d) Bank has to keep a database of digital signatures of all customers

6.21 The credit card issuing bank has to have the public keys of all customers in SET protocol as it has to
(a) Check the digital signature of customers
(b) Communicate with merchants
(c) Communicate with merchant's credit card company
(d) Certify their keys

6.22 One of the problems of using SET protocol is
(a) The acquirer does not know what was purchased
(b) The merchant does not know the credit card number
(c) All customers need a certified public key
(d) The bank does not know what was purchased

6.23 Dual digital signature is required to
(a) Settle disputes, if any, between a customer and a merchant
(b) Ensure authenticity of customer
(c) Ensure authenticity of acquirer
(d) Settle disputes between acquirer and credit card issuing bank

6.24 In order to check customer's digital signature, the merchant needs the
(i) Customer's private key
(ii) Customer's credit card number
(iii) Customer's public key certificate
(iv) Hash function used by the customer
    (a) (i), (ii)   (b) (ii), (iii)
    (c) (iii), (iv)   (d) (i), (iv)

6.25 The 9-digit MICR number of payee's bank is required by payer in
(a) Electronic Funds Transfer (EFT) only
(b) Electronic Clearing System (ECS) only
(c) Both EFT and ECS
(d) Neither EFT nor ECS

6.26 The 9-digit MICR number of payee's bank is required by ACH in
(a) EFT only   (b) ECS only
(c) Both EFT and ECS   (d) In neither EFT nor ECS

6.27 The function of ACH in EFT is to
(i) Maintain credit and debit balances of member banks
(ii) Sort cheques sent by payer's bank and send them to appropriate payee's bank
(iii) Send clearance instruction to payer's bank
(iv) Check payer's signature in cheques for authenticity
    (a) (i), (iii)   (b) (i), (ii), (iii)
    (c) (i), (ii), (iii), (iv)   (d) (i), (iii), (iv)

6.28 The function of ACH in ECS is to
(i) Maintain credit and debit balances of member banks
(ii) Sort cheques sent by payer's bank and send them to appropriate payee's bank

(iii) Send instructions to payee's bank to credit payee's account
(iv) Authenticate payer's instructions
    (a) (i), (ii)   (b) (i), (ii), (iii)
    (c) (i), (ii), (iii), (iv)   (d) (i), (iii), (iv)

6.29 The disadvantages of traditional ACH in cheque clearance from the point of view of payee is
  (i) Clearance delayed by at least a day
  (ii) Cheque can be dishonoured due to various reasons
  (iii) Payer can use cheque amount till it is cleared
  (iv) Transaction cost is high for outstation cheques
    (a) (i), (ii)   (b) (i), (ii), (iii)
    (c) (i), (ii), (iii), (iv)   (d) (i), (ii), (iv)

6.30 The advantages of ECS for a payee is
  (i) Amount is immediately credited
  (ii) No dishonouring of payment possible
  (iii) Payer can use ECS amount till it is cleared
  (iv) Transaction cost is high to payer
    (a) (i), (ii)   (b) (i), (ii), (iii)
    (c) (i), (ii), (iii), (iv)   (d) (i), (ii), (iv)

6.31 Electronic cheque payments will be mostly used in
  (a) C2C e-commerce   (b) C2B e-commerce
  (c) B2B e-commerce   (d) Interbank account reconciliation

6.32 To implement e-cheque payments
  (i) Both businesses must digitally sign e-cheque/deposit slip
  (ii) The payer businesses must digitally sign e-cheque/deposit slip
  (iii) ACH must be able to process e-cheques
  (iv) A standard format is needed for e-cheques
    (a) (i), (ii)   (b) (i), (ii), (iii)
    (c) (i), (ii), (iii), (iv)   (d) (iii), (iv)

6.33 In B2B e-cheque payment the
  (a) Payee encrypts e-cheque with his public key
  (b) Payer encrypts the e-cheque with payee's public key
  (c) Bank encrypts the e-cheque with its private key
  (d) ACH encrypts the e-cheque with its public key

6.34 In B2B e-commerce the vendor receives from the purchaser
  (a) Invoice reference with e-cheque digitally signed by purchaser
  (b) E-cheque digitally signed by purchaser
  (c) Invoice reference, e-cheque and vendor's public key certificate signed digitally by the purchaser
  (d) Invoice reference, e-cheque, purchaser's public key certificate are digitally signed and encrypted with vendor's public key

6.35 ECS payment in B2B e-commerce
  (a) Needs a standard e-cheque format
  (b) Does not need ACH
  (c) Needs public key certificates of all parties involved
  (d) Needs the payer to send to ACH an encrypted pay order encrypted with ACH's public key

6.36 In ECS payment in B2B e-commerce the
(i) Payer sends details of payee's account and bank code along with amount to be paid encrypted using ACH's public key
(ii) ACH sends instruction to credit payee's account with specified amount encrypted using the public key of payee's bank
(iii) ACH digitally signs its instruction to payee's bank
(iv) Payer digitally signs his or her instruction to ACH
 (a) (i), (ii)
 (b) (i), (ii), (iii)
 (c) (i), (ii), (ii), (iv)
 (d) (i), (iii), (iv)

6.37 Communication between ACH and banks will normally use
 (a) Internet
 (b) Intranet
 (c) Intranet using VPN
 (d) Intranet using leased lines

6.38 When a bank sends pay order to ACH in a secure envelope, we mean all relevant information is put together and
 (a) Digitally signed by the bank
 (b) Encrypted using bank's public key
 (c) Encrypted using bank's private key
 (d) Encrypted using ACH's public key

6.39 In electronic cash payment
 (a) A debit card payment system is used
 (b) A customer buys several electronic coins which are digitally signed by the bank which issued them
 (c) A credit card payment system is used
 (d) RSA cryptography is used in the transactions

6.40 In electronic cash payment
(i) A customer withdraws e-coins in various denominations signed by the bank
(ii) The bank has a database of issued coins
(iii) The bank has a database of spent coins
(iv) The bank cannot trace a customer
 (a) (i), (ii)
 (b) (i), (ii), (iii)
 (c) (i), (ii), (iii), (iv)
 (d) (ii), (iii), (iv)

6.41 In e-cash system described in this chapter, the vendor
 (a) Can exchange e-coins with other vendors
 (b) Can return e-coins as change to any customer
 (c) Can deposit e-coins with issuing bank and get it credited to its account
 (d) Need not have an account with the issuing bank

6.42 In e-cash system
(i) There is a limit to the amount of e-cash a bank will issue
(ii) The bank will issue any amount
(iii) E-cash is used only for small payments
(iv) Transaction cost is lower than in credit card
 (a) (i), (iii), (iv)
 (b) (iv), (iii)
 (c) (i), (iv)
 (d) (i), (iii)

6.43 In e-cash system several e-coin dispensing servers are used to
   (i) Improve system scalability
   (ii) Improve system reliability
   (iii) To some extent provide anonymity on spending habits of customers
   (iv) To allow return of e-change
      (a) (i), (ii)
      (b) (i), (ii), (iv)
      (c) (i), (ii), (iii)
      (d) (i), (ii), (iii), (iv)

6.44 In anonymous e-cash system
   (i) The customer mints e-coins of requisite denominations and assigns them random 50-digit serial numbers and blinds these
   (ii) The bank signs these e-coins without knowing the customer's identity
   (iii) Merchants accept e-coins as they are signed by the bank
   (iv) Double spending can be detected by the bank with some difficulty
      (a) (ii), (iii)
      (b) (iii), (iv)
      (c) (i), (ii), (iii), (iv)
      (d) (i), (iii), (iv)

6.45 In anonymous e-cash, the
   (a) Customer blinds e-coins with a random number
   (b) Bank blinds the e-coins before issuing
   (c) Bank is unable to sign blinded coins
   (d) Blinding factor used by customer is revealed to the bank

6.46 In anonymous e-cash, e-coins are blinded using the procedure
   (a) A random number is encrypted with bank's public key and multiplied by the e-coin serial number by the customer
   (b) The product of e-coin serial number and a blinding number is encrypted with bank's public key by the customer
   (c) A blinding random number is encrypted with bank's public key and the serial number with customer's public key and multiplied
   (d) A blinding random number is encrypted by the customer's private key and the serial number by bank's public key and multiplied

6.47 In anonymous e-cash
   (a) The bank mints e-coins and sends them to the customer for blinding
   (b) The customer deposits an amount with the bank and mints e-coins with random serial numbers and blinds them
   (c) A merchant issues blinded coins to a customer with bank's permission
   (d) The bank mints and blinds e-coins, signs and sends them to the customer

6.48 In the Net Bill's protocol for small payment for information goods the
   (i) Customer is charged only when the information is delivered
   (ii) Vendor is guaranteed payment when information is delivered
   (iii) Customer must have a certified credit card
   (iv) Customer must have a certified public key
      (a) (i), (ii)
      (b) (i), (ii), (iii), (iv)
      (c) (i), (ii), (iii)
      (d) (i), (ii), (iv)

6.49 In Net Bill's protocol for small payment for information goods
   (i) Key to decrypt information is sent to the customer by Net Bill only when there is enough amount in the customer's account

(ii) The vendor supplies the decryption key to Net Bill server when he or she receives payment
(iii) Checksum of encrypted information received by customer is attached to his or her payment order
(iv) The vendor does not encrypt information purchased by customer
    (a) (i), (ii)   (b) (i), (ii), (iii)
    (c) (i), (ii), (iii), (iv)   (c) (i), (ii), (iv)

6.50 In the Net Bill system the
    (a) Customer prepays some amount to Net Bill
    (b) Customer pays to Net Bill only after it sends decryption key
    (c) Customer accumulates several debits and pays Net Bill
    (d) Net Bill debits customer's debit card

6.51 The main disadvantage of e-cash payment using smart cards in e-commerce is
    (a) Smart cards should be physically sent to the bank for replenishment
    (b) PC should be equipped with a smart card reader
    (c) They are insecure
    (d) If the card is lost money is lost

6.52 In smart card cash payments
    (a) The bank should check double spending
    (b) The bank need not check card's authenticity
    (c) The merchant should check double spending
    (d) Once money is debited by the bank it cannot be respent

6.53 A payment gateway
    (i) Is a proxy which acts on behalf of a merchant to facilitate payment
    (ii) Can handle both credit card and electronic ECS payments
    (iii) Can handle e-cash payments
    (iv) Authenticates customers on behalf of merchants
    (a) (i), (ii)   (b) (i), (ii), (iii)
    (c) (i), (ii), (iii), (iv)   (d) (i), (ii), (iv)

6.54 Pay Pal is a payment service which
    (i) Is useful in C2C e-commerce
    (ii) Settles accounts among its members who maintain accounts with it
    (iii) Acts like an Internet based bank accepting deposits and payments
    (iv) Acts as an acquirer in credit card payments
    (a) (i), (ii)   (b) (i), (ii), (iii)
    (c) (i), (ii), (iv)   (d) (i), (ii), (iii), (iv)

CHAPTER 7

Structured Electronic Documents

LEARNING GOALS

In this chapter, we will learn:
1. Why standardization of format for electronic documents is essential for exchanging business documents in B2B e-commerce.
2. Electronic Data Interchange Standard and its drawbacks.
3. About the use of XML in electronic data interchange.
4. Basic ideas of good web site design.

7.1 INTRODUCTION

In Chapter 1, we briefly described the need for electronic documents in intranets and for exchange of documents between business partners. In this chapter, we will review and expand it to motivate the need for standardization of electronic documents in B2B e-commerce. We will first examine the standardization methods used in pre-e-commerce era. We will then examine how the document exchange method is evolving with the advent of the Internet and B2B e-commerce. Whereas e-document exchange is very important in B2B e-commerce, it is not so in B2C e-commerce. In B2C e-commerce, an attractive web site of a business (or an intermediary in the case of C2C e-commerce) which is easy to use and navigate by lay users of computers is of prime importance. We will discuss these aspects in the last part of this chapter.

7.2 ELECTRONIC DATA INTERCHANGE

In Chapter 1, we described the need for vendors and purchasers to have their own intranets and the need for the two businesses to be connected by a secure communication line to enable them to exchange e-documents. We will expand and revisit that example in this section.

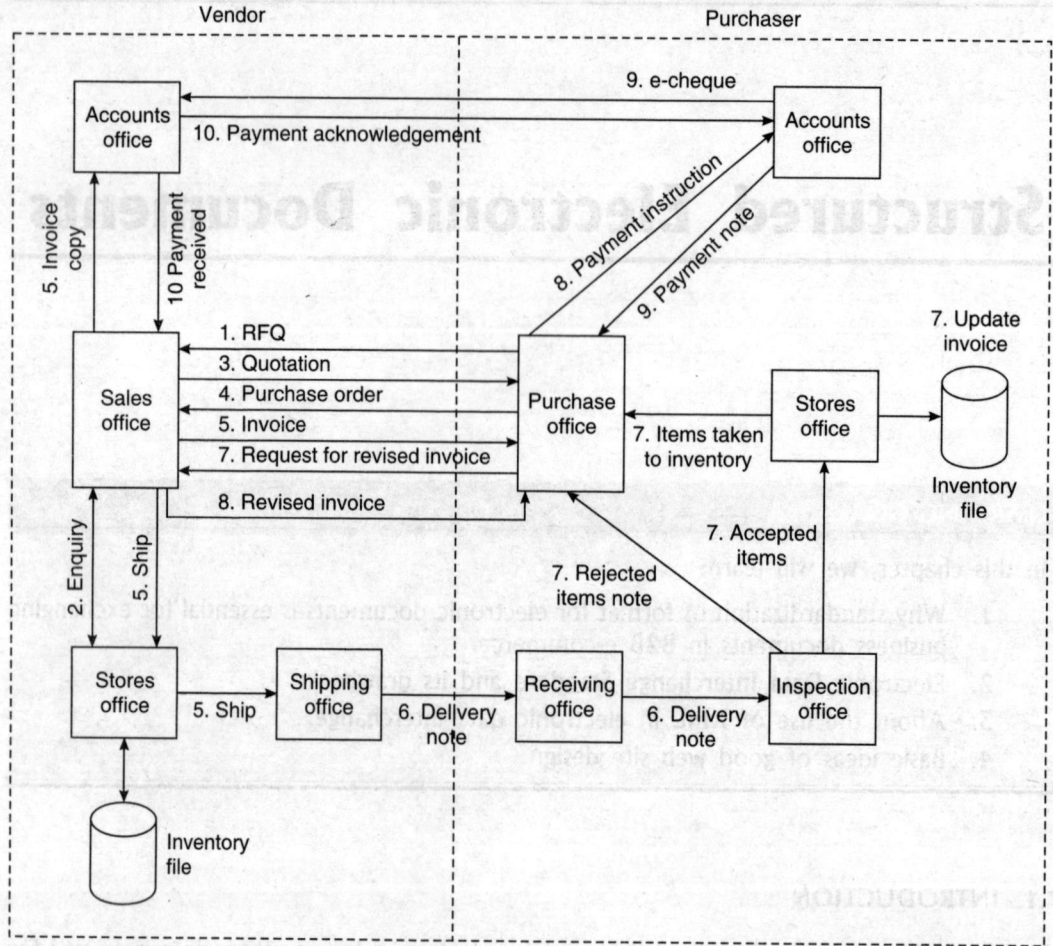

Figure 7.1 Document flow within and between a vendor and a purchaser offices.

Referring to Figure 7.1, we see that in a purchase operation two parties are involved: a vendor and a purchaser. The steps followed are:

1. The purchase office of the purchasing organization sends a Request For Quotation (RFQ) to a vendor's sales office.

2. The vendor's sales office enquires with its stores about the availability and quantity available of the item(s).
3. Based on the reply the vendor sends a quotation to the purchase office.
4. The purchase office examines the quotation and if OK sends a purchase order to the vendor.
5. The vendor sends an order acceptance and an invoice to the purchaser. It simultaneously sends a request to its stores office to supply the goods. The stores office in turn requests the shipping office to ship the items invoiced.
6. The shipping office sends the items along with a delivery note to the receiving office of the purchaser which forwards this along with the items received to the inspection office.
7. The inspection office examines items received (both quantity and quality) and sends a note on accepted items along with the items to the stores office. Simultaneously, it informs the purchase office about rejected items which is immediately communicated to the vendor's sales office with a request for a revised invoice. The stores office takes the items into stock and informs the purchase office.
8. The revised invoice is sent to the accounts office for payment to the vendor.
9. The accounts office sends an e-cheque to the vendor's accounts office along with the revised invoice and informs the purchase office.
10. The vendor's accounts office acknowledges receipt of the cheque to the purchaser and simultaneously informs its sales office.

The above procedure illustrates how a large number of documents have to flow among the offices of each of the organizations transacting business and also between the organizations. Before the advent of computing all these were paper documents which required enormous time and effort with a large number of clerks doing the processing. When computers came in these paper documents were converted to computer readable documents by manually entering all the data on terminals. Thus, organizations were keen to avoid this manual data entry which was expensive and prone to errors. There were two requirements:

1. Standardization of electronic format of documents within an organization to enable fast handling and processing of inter-office document flow.
2. Standardization of common documents normally exchanged between business partners and make them computer readable to eliminate manual entry of documents. Eliminating manual entry reduces cost and ensures accuracy.

As it happened, the electronic documents within each organization were standardized to ease the design of business data processing programs which are company specific. However, for exchange of documents between two organizations a standard had to be evolved. IBM which was a dominant computer vendor in the 1970s designed a standard for Electronic Data Interchange (EDI). This was modified by the American National Standards Institute and was published as ANSI X 12 EDI standard. Another standard was developed as an International standard by the United Nations Economic Commission for Europe and was called EDIFACT (Electronic Data Interchange Format for Administration Commerce and Transport). Currently EDIFACT is used by most countries outside USA whereas in USA, the ANSI X 12 EDI standard is used.

There are several definitions for EDI. EDIFACT standard definition is: "Electronic Data Interchange (EDI) is the interchange of standard formatted data between computer application systems of trading partners without manual intervention."

In other words, the internal document structure of a Business A has to be transformed to the EDIFACT standard of that document before sending it to its trading partner Business B. Business B's computer should receive this, interpret it and transform it to its own internal format (See Figure 7.2). Similarly, B's internal document formats have to be translated to EDIFACT and sent to A's computer which should in turn convert them to their own internal format. You may wonder why the businesses should not adopt EDIFACT as internal standard also. However, EDIFACT standard was evolved before the advent of Internet and the current generation of high speed communication and computing. Thus, they tried to minimize the size of documents and make them simpler for computer processing. This makes these forms difficult to use for local processing as it is not understandable to humans. Internal documents of businesses are designed to be understandable and to simplify data processing programs.

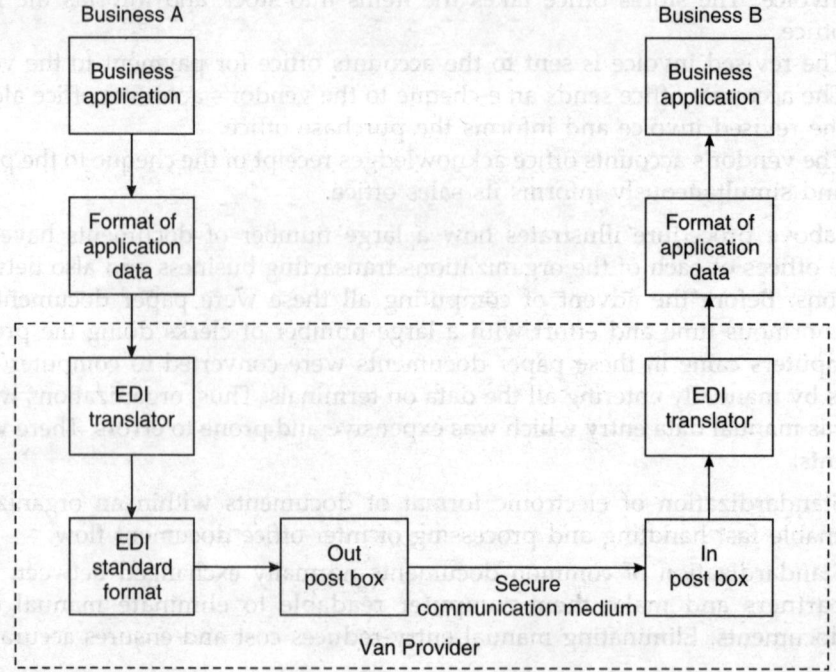

Figure 7.2 Steps followed when EDI format is used (VAN provider's services enclosed within broken lines).

The common structure of EDI documents contain:

1. *Transaction set* which is equivalent to a business document such as purchase order, invoice etc. Each transaction has several data segments. There are several hundred transaction sets defined in the standard.
2. *Data segments* which is a group of data elements that together convey some information, for example, a purchase order line, company name and address.
3. *Data elements* which are individual fields such as price of an item, quantity, etc.

A sample EDI document for a book purchase order is given in Table 7.1.

Table 7.1 A sample EDI document for a book purchase order

EDIFACT form	Meaning
UNH000002 + ORDERS: DD96A UN:EAN 008	Header
BGM + 220 – XY0342-9'	Order No. XY0342
DTM -137 – 20080415:102'	Message Date YYYYMMDD
NAD + BY STANDARD BOOK COMPANY: 2+ A.B.C. 120 AD, BANGALORE++560022'	Purchaser's name and address
REF + API:SBC7432'	Purchaser's code
NAD+SU+++PHINC'	Vendor's name
CUX + 2: USD:9'	Order currency

Observe that the document format is difficult to understand. It is meant to be interpreted only by computers.

Developing data processing systems with business rules which are unique to each organization is difficult if EDIFACT is used. Thus, the most common method adopted by several companies is to have a third party called Value Added Network Service provider to handle EDIFACT data formats. This is not essential but it is convenient (see Figure 7.2).

VAN provider's responsibility is to:

1. Provide secure and stable communication among business partners.
2. Translate internal formats of each of its members to a standard EDI format (EDIFACT or ANSI X 12) and vice versa.
3. Provide 7 days a week, 24 hours per day service so that documents may be received at any time and be stored and forwarded to the appropriate addressee.
4. Transactions roll back to support uncommitted transactions.
5. Provide audit trails for documents received to settle any disputes which may arise between businesses.
6. Provide a good customer support.

EDIFACT and other standards define formats for a range of business documents such as purchase order, tenders, bill of ladings and invoices. The number is enormous and most businesses find it convenient to employ a VAN provider. However, VAN providers are expensive and many Small and Medium Businesses (SMBs) find them beyond their budgets. Thus, EDI adoption has been sluggish with less than 10% of businesses adopting them. Only when compelled like in international trade (India's Customs Department insists on EDIFACT for bill of lading) or by big businesses (such as Wal Mart) do companies adopt EDI standard.

7.2.1 Transporting EDI Formatted Data

So far we have not discussed in detail how EDI documents are transported between business partners. In the early years of EDI (1970–90) the documents were transported physically using magnetic tapes in which EDI data were recorded. Later floppy disks were used. When communication systems improved, leased lines were used by VANs. With the advent of the Internet the entire scenario has changed. Now there are many alternatives to transport EDI data. One may send EDI data by e-mail using S/MIME standard. One may also send EDI files using ftp after encrypting the files using 3DES/RSA combination. Another alternative is to use the World Wide Web infrastructure and use https. The main point is that EDI is independent of the communication protocol and the physical transport layer. The physical transport layer may be leased lines, extranets using VPN and even the Internet if security is ensured.

One of the services provided by VANs is to maintain leased lines and electronic post boxes between business partners. With the advent of the Internet the communication function of VANs has lost its importance. Small and Medium Businesses can adopt EDI using the services of EDI gateway programming services which will provide the following services:

1. Translation programs to convert company specific forms to EDI standard data format
2. Translate EDI standard data format received from business partners to company specific formats
3. Provide audit trails of sent and received documents ensuring non-repudiation
4. Ensure security of data transported among business partners
5. Transaction roll back to support uncommitted transactions
6. Overall customer support

7.3 EDI AND XML

The Electronic Data Interchange standard is primarily aimed at sending data in a form which is readable and processable by computers. The aim of EDI is restricted. In general, when we state that data should be represented digitally there are several requirements which ideally should be met. They are:

- Data should be readable by computers.
- Data should be in a form which will be easy to process by computers.
- Data should be storable with appropriate structure to enable search and retrieval.
- It should be displayable in a form understood by users of data and be printable.
- It should be possible to disseminate the data easily, preferably in digital form for display/printing at remote locations.

If we look at a variety of documents they have three basic characteristics. They are:

- *Content*—The matter in the document. For example, the content of a letter in its main body conveying desired information.
- *Structure*—The document type and the organization of elements within it. For example, a business letter has the structure shown in Figure 7.3. The structure of documents differs based on their nature. For example, the structure of a recipe is quite different from that of a business letter. The structure defines the kind of elements a document contains and the order in which they can occur.
- *Presentation*—The format in which the information is presented to the reader. It may differ depending on the medium in which it is presented. For example, a

letter typed on paper usually follows a company specific format. The same letter sent by e-mail will have a different format. The font used, the size of characters, etc., is variable based on one's taste.

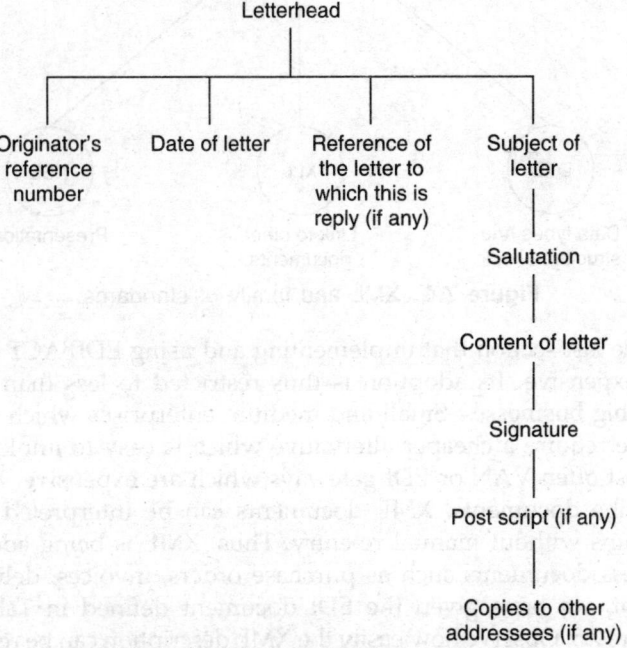

Figure 7.3 Structure of a business letter.

The main idea of XML is the realization that significant benefits accrue in computer processing of documents if these three aspects of a document are kept separate. In paper based documents such as purchase order, invoice etc., the content, structure and format are all mixed up.

XML is primarily a language to model documents of diverse types. It has syntax and semantic rules just like any other programming language. Its structure can be checked by what is known as Document Type Definition (DTD) which specifies the data types of variables used in XML and also whether XML is well structured. Data types of variables are essential to process a document using a computer. How a document looks when printed or seen on a browser is relegated to another standard known as XML Style Language (XSL). Thus, by changing XSL the presentation style of an XML document may be changed. Another feature of XML is the facility to link it to other documents. XML has several new features for linking compared to HTML. In order to allow links to other documents there is another standard known as Extensible Link Language. Thus, XML is a family of standards consisting of XML, DTD, XSL and XLL (See Figure 7.4). We will briefly describe this in the next section. There are plenty of books on XML which give in detail all features of XML. Hence, we will not describe them in this book. We give in the set of references several books on XML.

158 *Essentials of E-Commerce Technology*

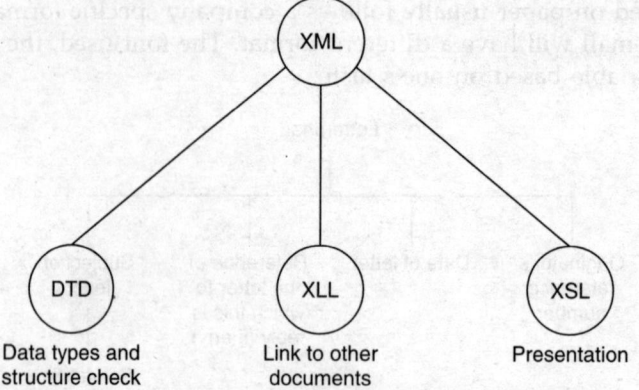

Figure 7.4 XML and family of standards.

We saw in the last section that implementing and using EDIFACT or ANSI based EDI is inflexible and expensive. Its adoption is thus restricted to less than 10% of businesses which are mostly big businesses. Small and medium enterprises which want to participate in B2B e-commerce require a cheaper alternative which is easy to implement and use. Use of EDI implies most often VAN or EDI gateways which are expensive. XML is an excellent language to describe documents. XML documents can be interpreted and processed by application programs without manual re-entry. Thus, XML is being adopted by SMBs for exchanging business documents such as purchase orders, invoices, delivery notes, etc. To illustrate this point, we have given the EDI document defined in Table 7.1 as an XML description in Figure 7.5. Observe how easily the XML description can be read and understood. When a business uses XML to describe business documents, it also gives a set of statements which define the data types of elements used in the XML program. This is called a *Document Type Definition* (DTD). This is published in the company's web site so that any application program wanting to use the XML document can use the DTD to interpret the document. The DTD corresponding to the XML description of Figure 7.5 is given as Figure 7.6. In this definition, PCDATA means that the element contains a text. There are other keywords used in DTD, which we will not discuss in this book. The file where DTD is available (name) is given at the beginning of the XML program of Figure 7.5.

DTD is also used to check the correctness of documents which use XML definition. For example, if PIN code is essential in an address this can be specified in the DTD. If it is missing in a document, an error message appears.

```
<?XML Version = "1.0">
<!DOCTYPE order SYSTEM"order.dtd">
<order>
    <order-no>/XY0342</order-no>
    <date>
        <year>2008</year>
        <month> 04</month>
        <day>15</day>
    </date>
    <purchaser>
```

```
        <name> Standard Book Company </name>
        <address>
            <street> 2A.B.C. Road </street>
            <city> Bangalore </city>
            <pin-code> 560022 </pin-code>
        </address>
        <purchaser-ID> SBC7432 </purchaser-ID>
    </purchaser>
    <supplier>
        <name> PHINC </name>
    </supplier>
    <currency-type> US Dollars</currency-type>
</order>
```

Figure 7.5 XML definition of book purchase order given in Table 7.1 in EDIFACT notation.

```
<!.. This is a comment with delimiters ..>
<! ELEMENT order (entry +)>
<!.. order is top-level element and is a list of 1 or more entries..>
<!.. an entry is an order-no followed by date, purchaser, supplier and
currency type ..>
<! ELEMENT order-no (# PCDATA)>
<.. #PCDATA means a character string ..>
<! ELEMENT date (year, month, day)>
<! ELEMENT year (# PCDATA)>
<! ELEMENT month (# PCDATA)>
<! ELEMENT day (# PCDATA)>
<! ELEMENT purchaser (name, address, purchaser-ID)>
<! ELEMENT name (# PCDATA)>
<! ELEMENT address (street, city, pin-code)>
<! ELEMENT street (# PCDATA)>
<! ELEMENT city (# PCDATA)>
<! ELEMENT pin-code (# PCDATA)>
<! ELEMENT purchaser-ID (# PCDATA)>
<! ELEMENT supplier (name)>
<! ELEMENT name (# PCDATA)>
<! ELEMENT currency-type (# PCDATA)>
```

Figure 7.6 Document type definition for order.

As we saw businesses design documents to suit their business rules. Thus, documents of two businesses may have different formats. When a business sends, say, an XML-based purchase order, it can be interpreted and converted to that appropriate for the receiving business as XML is easy to interpret and process. However, it requires extra programming effort. Thus, efforts are underway for standardizing common business documents such as purchase order and invoice using XML. One reasonably successful effort has led to open financial exchange initiative which has standardized XML documents for on-line banking. We end this section with a comparison of various structured electronic documents (Table 7.2).

Table 7.2 Comparison of structured documents

	Flexibility	Implementation	Cost	Communication channel
Traditional EDI	Poor	Difficult	High	Private VAN
EDI gateway	Poor	Easier	Moderate	Internet/VPN
XML based EDI	Very good	Easy	Low	Internet/VPN

7.4 XML AND HTML

We described HTML in Chapter 4 as a language for designing web pages. A question you may have is the role of XML compared to HTML. Both HTML and XML have SGML as their parent. In other words, the idea of markup is common to both of them. While the role of HTML is primarily on document formatting and presentation, that of XML is to allow document processing. As we saw XML has separate standards, namely, XLL and XSL while HTML's strength is the simplicity of linking and formatting. All browsers support HTML. XML support is now available on browsers. XML with XLL and XSL can be used to design web pages also.

Both HTML and XML will continue to be used in their own domains of strength and will cooperate in several applications.

XML has several applications. A few of them are:

1. *On-line banking*: A standard XML document format known as *open financial exchange initiative* is used for obtaining information such as bank statements and for sending standing instructions to banks.
2. *Software distribution*: Software updates are distributed periodically using XML.
3. *Scientific publishing*: Markup languages modelled using the ideas of XML are used to describe documents of interest to scientists. For example, CML is a markup language to describe complex chemical molecules. Mathematical markup language MML is used to describe complex mathematical formulae.

We briefly mentioned about XLL and XSL. We will very briefly describe how documents are linked in XML and about style sheets for formatting XML documents. From the point of view of document structuring and exchange, these are not very important. Only for display they are necessary. Thus, our discussion will be superficial to give you a flavour.

XML uses reserved attribute xlink to allow elements to signal that they are link resources. The value of the attribute is used to specify the type of link element. The link identifier A used in HTML can be used and is known as a simple link in XML.

```
<A>
    xlink:type = "simple"
      xlink:href = "http://www.xxfurniture.com">
    Click here for details of products of xxfurniture
</A>
```

There is another link called extended link which allows use of pointers to locate specific item within a web page.

7.4.1 XSL Style Sheet

The objective of XSL is to provide, easy to use style sheet syntax for expressing how XML documents should be rendered. Using XSL description one should be able to use a computer program to convert it into HTML, PDF, or any other formatting file. XSL is based on Document Style and Semantics Specification Language (DSSSL), an international standard style sheet language for formatting SGML. DSSSL is big and powerful. Thus, XSL is an abbreviated version. It borrows several ideas from DSSSL but its syntax is based on XML. XSL has also recognized the importance of generating HTML from XML (See Figure 7.7).

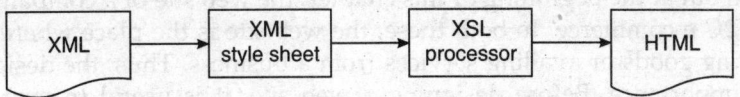

Figure 7.7 Generating HTML from XML using XSL.

Let us take a simple example just for illustration. An instance of a document described using XML is given in Figure 7.8.

```
<book>
    <title> Electronic Commerce </title>
    <author> V.Rajaraman </author>
    <para>
This book gives the essentials of e-commerce technology
    </para>
</book>
```

Figure 7.8 XML description of a book.

An XSL transformation processor to render this using HTML would examine the root element of the document, namely, <book> and the rules to be applied to this by template matching of this root.

Thus, it will look like

```
<xsl : template match = "book">
    <! Describe how book should be formatted using html>
</xsl : template>
```

We give in Figure 7.9 a simple part of a style sheet which gives the output shown below it (Ignore <u>exact syntax</u>)

<title> in <book> = <h₁><CENTRE> style = " font-family:`Times New Roman';font-size = 24pt;font-weight=bold"</CENTRE>
</h1>
<author> in <book> = <h₂><CENTRE> style = "font-family:`Courier';font-size = 18 pt;font-weight=bold"</CENTRE> </h₂>
<p> in <book> = style = "font family:`san serif';font-size = 10 pt" </p>

Electronic Commerce

V. Rajaraman

This book gives the essentials of e-commerce technology.

Figure 7.9 Rendering with style sheet.

We have given a very naïve description as XSL is not important in processing XML documents. Those interested in learning about XSL in detail should read the numerous references given at the end of this book. The primary intention of this description is for you to understand that rendering is separated from XML and several possible renderings are possible, the most popular being HTML.

7.5 BASICS OF WEB SITE DESIGN

As we pointed out at the beginning of this chapter, the web site of a company is important in B2C and C2C e-commerce. In both these, the web site is the place where the customer looks for buying goods or availing services from a business. Thus, the design of web site is of crucial importance. Before designing a web site, it is useful to visit a number of existing web sites to see what are the good and bad aspects in their design. Some sites such as amazon.com, a pioneering book store, has refined their web site over a period of time based on experience. Thus, it is essential to visit these. The site homestead.com gives a number of samples of well-designed home pages, namely, the first pages of the websites of a number of small businesses. It is a good idea to look at them to understand the importance of simple user-friendly designs.

We will briefly present in this section a set of simple rules which should be remembered when designing a web site of a business or a service provider. The home page of a web site is the first page a customer will see when he or she clicks the URL of that site. Thus, it should be attractive and not cluttered with too much detail. Fancy graphics may look good but it is information and a quick response to his requests that a customer values. If it takes too long to download the graphics, it will be counter productive. Fast access to the home page is a very important requirement.

Almost all businesses and professionals now have a web site. We can classify them into the following categories:

1. Shops which have a physical presence and which have a web site as a means of advertising their goods and expect people to visit and buy at their shops, clothing store, etc.
2. E-shops which have no physical presence but sell goods and send them by mail/courier. Typical shops of this type are bookstores, music stores, video stores, etc.
3. Professional services such as group practice of doctors, lawyers, accountants, etc.
4. Sites for booking tickets to theatres, trains, airlines, etc. Hotels and travel agents are in the same category.
5. Auction sites where customers log on to buy and sell goods in C2C e-commerce.
6. Big businesses which primarily use web sites to publicize their products or services, catalogues, news and other important information. They are not normally in retail sales.

Each category requires web site design appropriate for its line of business or service. There are, however, some requirements which are common to all of them. They are:

1. A web site is an important marketing tool. Thus, the home page should be attractively designed and briefly describe the unique business of the company.

2. The title line is often used to get the keywords by search engines. Thus, this line must include keywords appropriate to the type of business. For example, a bakery may have a title "Best Bakery Products in Town".
3. The home page must not have a cluttered look. It must have only a set of important items in the main menu (around 7) for a visitor to click to get more information. An unique feature of a home page is its hyperlinks. This fact should be used to navigate to other web pages when a menu item in the home page is clicked. In succeeding pages one may give further sub-menus.
4. Use of too many colours is distracting. An aesthetically pleasing colour combination should be used.
5. Periodically the web site should be updated. The web site should always be up-to-date.
6. There should not be a menu selection which when selected is either blank or says "under development".
7. Lastly it will be useful to track customer behaviour (when he or she visits a site) using cookies if permitted by a customer.

A generic home page of a web site is shown in Figure 7.10. This gives the major menu selections which should be common to all businesses. The following points should be observed:

1. Menu selections are limited.
2. When there are several products or services the most important 2 or 3 are in the main menu. Clicking any of them leads to other set of related products or services which can be detailed in subsequent pages.

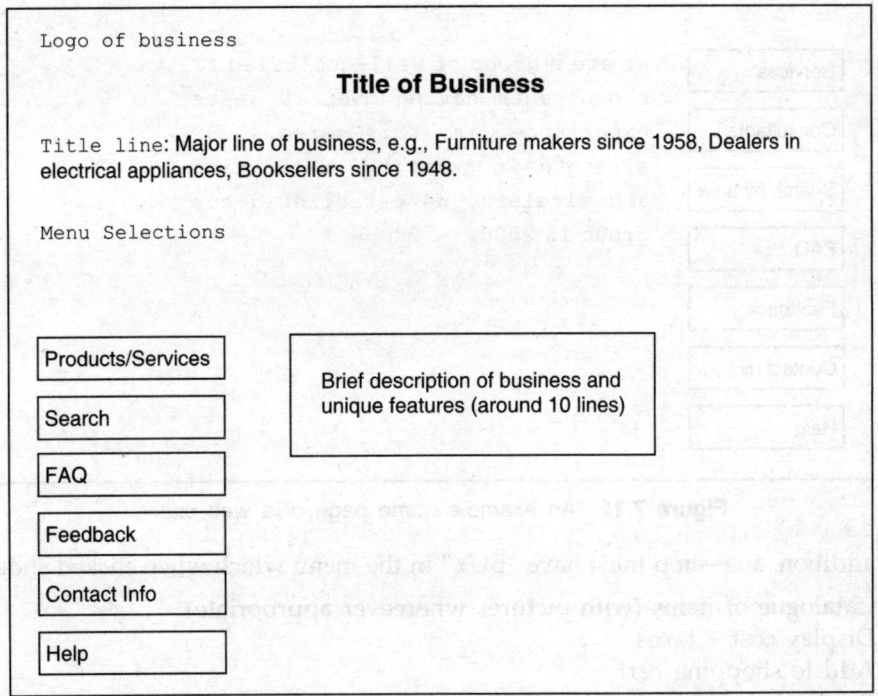

Figure 7.10 A generic home page of a web site.

3. There is a search facility. For example, a bookstore must also have facilities to search by author or subject. A restaurant should allow for search in their menu for specific items. A reservation system for airlines should allow search for schedules and available tickets.
4. A set of frequently asked questions and their replies relevant to the business or service should be included.
5. Provision must be there to obtain customer feedback on their satisfaction or otherwise of the service provided and or suggestions to improve the web site of the business.
6. Appropriate contact information such as address, e-mail id, phone number and fax number must be provided. Businesses such as restaurants, furniture shops, medical centres, etc., should also give a location map to allow customers to visit them.
7. Lastly, a help button to aid customers to navigate the site is useful.

An example of a home page is given in Figure 7.11.

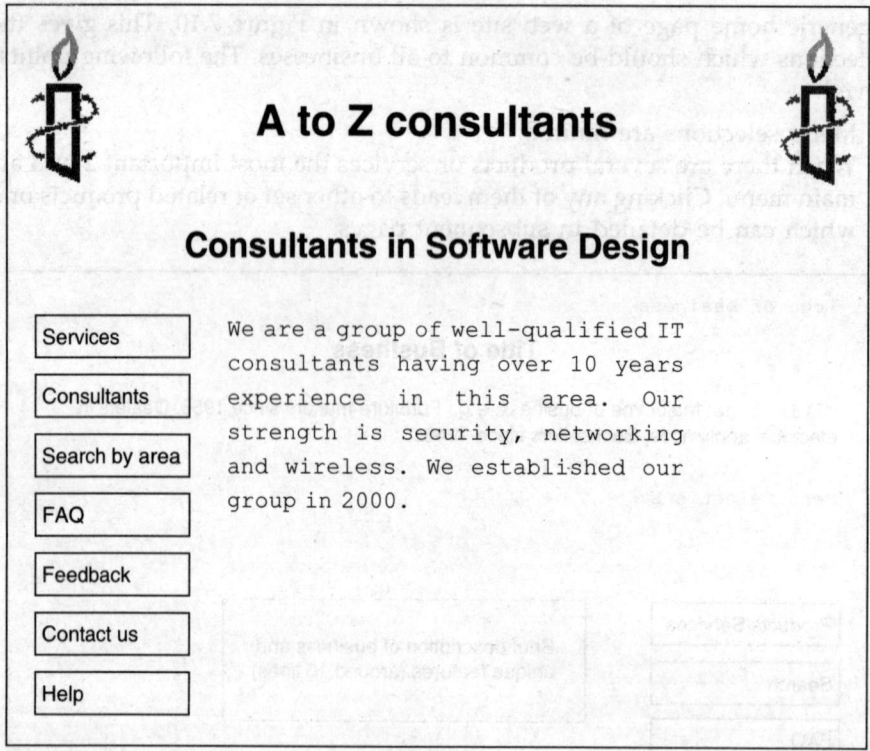

Figure 7.11 An example home page of a web site.

In addition, an e-shop must have "BUY" in the menu which when clicked should give

- Catalogue of items (with pictures whereever appropriate)
- Display cost + taxes
- Add to shopping cart
- Payment options
- Shipping address

- Shipping cost
- Billing address (if appropriate)
- Delivery details (number of days, mode of shipment)

A large business would have:
- Product catalogue
- Board of Directors
- Paid-up capital
- Distributors of products

and other relevant information.

A ticket booking site of airlines/railways/buses must have:
- Schedule of flights/trains/buses between cities where they operate
- Available tickets and cost in various classes
- Seat selection
- Booking option
- Payment normally by credit card on line or on delivery
- E-ticket printing
- Special request (if any)
- Contact address/phone number of traveller

We have not discussed the security of web sites as it has already been discussed in detail. Normally a business will provide the requirements to be met and it should be used to design a site appropriate for it.

There are many professional companies primarily in the design of web sites, maintaining them and updating them. They also host the web page 24 × 7 and are used by many small and medium businesses.

SUMMARY

1. In B2B e-commerce the participating businesses require exchange of documents such as purchase order, invoice, delivery note, etc. It is desirable if there is a standard followed for these common documents to eliminate all manual inputs. Electronic Document Interchange (EDI) format has been standardized since 1975 and used by big businesses.

2. EDI is defined as interchange of standard formatted data between computer application systems of trading partners without manual intervention.

3. All businesses have internal formats for documents to suit their business rules which do not follow EDI standard. EDI standard has been designed for machine processing and not for user understanding. Thus, there is a need to convert internal document format to EDI before forwarding to another business.

4. Value-Added Network providers translate business specific format to EDI and vice versa. They also provide 24 × 7 service, secure electronic transmission of data and audit trails. Thus, most businesses use VAN services when they use EDI. As VAN services are expensive only around 10% of businesses have adopted EDI for B2B e-commerce.

5. With the advent of the Internet and World Wide Web, secure electronic transmission is cheaper using this infrastructure. Businesses may thus outsource only EDI conversion job and use VPN or leased lines for their extranets. These services are called EDI gateway services.
6. Currently XML has emerged as a standard for structuring electronic documents. XML neatly divides the properties of documents as: content, structure and presentation. XML is used to structure content, DTD to define data types of elements, XLL to specify links to other documents and XSL for formatting.
7. The main advantage of XML is that it is designed with tags which aid in understanding the nature of a document. DTD accompanying XML specifies the data types of elements which allow XML documents to be processed.
8. A business sending an electronic document to another business can use XML to structure it and publish in its website DTD corresponding to this document. The receiving business can interpret it and write a program to convert it to its own internal form.
9. XML is designed for the World Wide Web. Thus, it is a cheaper alternative to traditional EDI and can use the Internet infrastructure.
10. Thus, businesses are adopting XML to exchange documents in B2B e-commerce.
11. Using XML along with XLL and XSL allows web pages to be designed. However, HTML is still the preferred language to design web sites due to its simplicity and universal availability with all browsers.
12. In B2C and C2C e-commerce the most important component of their website is the home page which is used to publicize the business and allow customers to buy their products or services.
13. Home pages are important not only for e-stores, but also for service providers such as law firms, medical centres and accounts firms.
14. Care must be taken to design a web site with the home page giving an uncluttered look with a small number of menu choices. The hypertext facility should be used to navigate the web site using submenus.
15. The design of web sites depends on the requirements of a business and is often outsourced to a professional web design and website hosting company.

EXERCISES

7.1 Why is there a need to standardize electronic documents in B2B e-commerce?
7.2 What problems will be encountered if businesses have to enter paper documents manually?
7.3 What is EDI? When was it first proposed?
7.4 What is EDIFACT? Who standardized EDIFACT?
7.5 Give the EDIFACT definition of EDI.
7.6 What are the drawbacks of EDIFACT specified EDI document? Can they be used as internal electronic documents of businesses? If no, explain why?
7.7 What is a VAN? What services does a VAN provide?

7.8 Why has EDI not been adopted by a large number of businesses?
7.9 What is the structure of an EDI document?
7.10 Is it essential for EDI formatted data to be sent only through a VAN? Can the Internet be used? If yes, how?
7.11 What is an EDI gateway? What services are provided by such a gateway? How is it different from a VAN?
7.12 What is XML? In what ways is it superior to HTML?
7.13 Give an XML definition of an invoice. Give its DTD definition.
7.14 What do you understand by content, structure and format? Give the structure of a text book. (*Hint*: Use the idea given in developing the structure of a business letter in the book).
7.15 Give the structure of an e-mail.
7.16 What are the roles of XML, DTD, XLL and XSL in describing documents?
7.17 Can XML be used for EDI? If so, how?
7.18 Are XLL and XSL used in EDI? Are they necessary?
7.19 When is it necessary to use XSL and XLL along with XML?
7.20 What are the strengths of HTML? What are the strengths of XML? Compare the features of XML and HTML.
7.21 Enumerate a set of rules which should be followed to design a good web site.
7.22 Give a figure showing the layout of the home page of a small restaurant's web site.
7.23 Look at the web site of `IndiaPlaza.com`. Enumerate the good and bad points of this site.
7.24 Look at the web site of `phindia.com`. Enumerate the good and bad points of this site.
7.25 What are the important requirements of an e-shopping site?
7.26 What are the important requirements of a ticket booking site? Take the example of irctc.com website of the Indian Railways?
7.27 What are the important requirements of a C2C web site?

OBJECTIVE QUESTIONS

Each question has four possible answers. Pick the most appropriate answer.

7.1 Documents exchanged among various departments of a business connected by an intranet must
  (i) Have a standard format appropriate for the business rules used by the business
  (ii) Use specially designed paper forms
  (iii) Be readable directly by computers
  (iv) Follow EDI standards
    (a) (i), (iv)
    (b) (i), (ii)
    (c) (i), (iii)
    (d) (ii), (iv)

7.2 To avoid manual data entry of documents in B2B e-commerce
   (a) Documents entered on standard form should be automatically read using a scanner
   (b) Use extranet to exchange documents
   (c) Use the Internet to exchange documents
   (d) Use standard electronic formats for documents exchanged between businesses

7.3 The expansion of EDI is
   (a) Electronic Document Interchange
   (b) Electronic Data Interchange
   (c) Exchange Data via Intranet
   (d) Electronic Data using Internet

7.4 The two major standards used in EDI are
   (i) EDIFACT
   (ii) ANSI X 12
   (iii) ANSI EDI
   (iv) ISOEDI
      (a) (i), (ii)
      (b) (iii), (iv)
      (c) (ii), (iii)
      (d) (i), (iv)

7.5 Use of EDI implies
   (i) Data formats are standardized for various business documents
   (ii) Data is transferred electronically between application systems of trading partners
   (iii) There is no manual data entry
   (iv) Use of VAN
      (a) (i), (ii), (iii), (iv)
      (b) (i), (iii), (iv)
      (c) (i), (ii), (iv)
      (d) (i), (ii), (iii)

7.6 EDI is used most in
   (a) B2C e-commerce
   (b) Commerce using Internet
   (c) C2C e-commerce
   (d) B2B e-commerce

7.7 EDI requires
   (a) Representation of common business documents in computer readable form
   (b) Data entry operators by business which receive the documents
   (c) Special value-added networks
   (d) Special hardware to interpret EDI documents

7.8 EDI standards are
   (a) Used only by a small number of big businesses
   (b) Still being evolved
   (c) Universally used in B2B e-commerce
   (d) Not used as it is very expensive

7.9 EDIFACT is a standard for
   (a) Representing e-mail among businesses
   (b) For ftp in B2B e-commerce

(c) Protocol used in B2B e-commerce
(d) Representing business forms used in B2B e-commerce

7.10 EDIFACT standard was developed by
 (a) American National Standards Institute
 (b) International Standards Organization
 (c) United Nations Economic Commission for Europe
 (d) European Common Market

7.11 In B2B e-commerce
 (i) Cooperating businesses should agree on EDI standard to be used
 (ii) Programs must be developed to translate EDI forms to forms accepted by application programs
 (iii) The method of transmitting/receiving data should be mutually agreed
 (iv) The intranet should be used
     (a) (i), (iii), (iv)   (b) (i), (ii), (iv)
     (c) (i), (ii), (iii), (iv)   (d) (i), (ii), (iii)

7.12 Value-added networks
 (i) Translate EDI to business specific forms and vice versa
 (ii) Provide secure networks with post boxes for subscribers to deposit their electronic documents
 (iii) Operate 24 hours a day, 7 days a week
 (iv) Guarantee delivery, audit trails, security and non-repudiation
     (a) (i), (iii), (iv)   (b) (i), (ii), (iii)
     (c) (ii), (iii), (iv)   (d) (i), (ii), (iii), (iv)

7.13 EDIFACT standard
 (a) Is not universally available
 (b) Defines several hundred transaction sets for various business forms
 (c) Defines a transport level protocol
 (d) Is difficult to use

7.14 EDI over the Internet may use
 (a) S/MIME to attach EDI forms to e-mail messages
 (b) SGML to send EDI forms
 (c) Normal ftp
 (d) HTTP protocol

7.15 To transmit EDI businesses may employ
 (i) VAN   (ii) VPN
 (iii) Leased lines
 (iv) The Internet with no special requirements
     (a) (i), (ii), (iv)   (b) (i), (iii), (iv)
     (c) (i), (ii), (iii)   (d) (i), (ii)

7.16 Value-added networks
 (a) Are essential for using EDI
 (b) Are an expensive option for small businesses
 (c) Use the Internet
 (d) Use their own EDI standard

7.17 EDI gateways are
(a) Gateways to send EDI documents
(b) Providers of EDI programming services to businesses
(c) Systems to send secure EDI messages
(d) Electronic transmission media which transmit secure EDI messages

7.18 EDI gateway services provide
(i) Translation of EDI formats to business specific formats and vice versa
(ii) Audit trails of documents exchanged between member businesses
(iii) Network for secure exchange of documents between members
(iv) 24 × 7 mail boxes in leased networks maintained by them
    (a) (i), (ii)   (b) (i), (ii), (iv)
    (c) (i), (ii), (iii)   (d) (i), (ii), (iii), (iv)

7.19 When we say data is represented in electronic form, we normally mean that it
(i) Is digital and readable by computers
(ii) Can be interpreted by programs
(iii) Can be transmitted using the Internet
(iv) Is always accurate
    (a) (i), (ii), (iii)   (b) (i), (ii), (iii), (iv)
    (c) (i), (ii)   (d) (i), (ii), (iv)

7.20 The basic characteristics of documents are
(i) Content
(ii) Structure
(iii) Format to enable presentation
(iv) Links to other documents
    (a) (i), (ii), (iii), (iv)   (b) (i), (ii), (iii)
    (c) (i), (ii), (iv)   (d) (ii), (iii), (iv)

7.21 The expansion of XML is
(a) Extra Markup Language
(b) Expanded Machine Language
(c) Extended Markup Language
(d) Exhaustive Markup Language

7.22 The objectives of XML are as follows:
(i) Allow processing of data stored in web pages
(ii) Allow users to design their own document types appropriate for the application
(iii) To provide features of HTML
(iv) To eliminate use of HTML
    (a) (i), (ii), (iii)   (b) (i), (ii)
    (c) (i), (ii), (iv)   (d) (iii), (iv)

7.23 XML
(a) Facilitate understanding of the type of a document
(b) Provides good formatting of documents
(c) Is an improvement over HTML
(d) Are very complicated and exhaustive

7.24 XML definition of a document consists of
   (i) Document Type Definition
   (ii) Defining a document with meaningful user-defined tags
   (iii) Formatting information
   (iv) XML description
      (a) (i), (iii)   (b) (iii), (iv)
      (c) (i), (ii)   (d) (i), (ii), (iii)

7.25 Links to XML specifications are provided by a special language known as
   (a) HTML link < A LINK>
   (b) XML link < A LINK>
   (c) XSL
   (d) XLL

7.26 XML documents can be displayed using
   (a) Cascaded Style Sheets
   (b) HTML
   (c) Extended Style Language
   (d) Extended Cascaded Style Sheets

7.27 XSL is translated by an XSL processor giving
   (a) HTML document
   (b) XML document
   (c) XLL document
   (d) Cascaded Style Sheets

7.28 XLL and XSL are used along with XML to
   (a) Replace HTML as a standard for web pages
   (b) Be able to display XML documents with hyperlinks on the World Wide Web with appropriate formats
   (c) Render XML documents on a display
   (d) Combine XML and html

7.29 Some of the important characteristics of the home page of a web site are:
   (i) Accessing it must be fast
   (ii) It must have an uncluttered look with a small number of menu items to choose from
   (iii) It must have plenty of graphics
   (iv) Must not use too many colours
      (a) (i), (ii)   (b) (i), (ii), (iv)
      (c) (i), (ii), (iii)   (d) (i), (ii), (iii), (iv)

7.30 The number of choices in the menu on the first page of a web site must be
   (a) At least 25   (b) Not more than 4
   (c) Any number   (d) Around 7

7.31 Shops with a physical presence and services such as doctors, accountants and lawyers must provide in their home page
   (a) Contact details including a site map of their address
   (b) Only telephone number
   (c) Details of prices of items or services
   (d) References of other customers

7.32 An e-shop must provide in their web site
   (i) Catalogue of items available
   (ii) Option to buy an item and add it to shopping cart
   (iii) Shipping address
   (iv) Payment options including personal cheques
       (a) (i), (ii), (iii), (iv)   (b) (i), (ii), (iii)
       (c) (i), (ii), (iv)   (d) (ii), (iii), (iv)

CHAPTER 8

M-Commerce

> **LEARNING GOALS**
>
> In this chapter, we will learn:
> 1. Emerging e-commerce applications using mobile hand-held devices.
> 2. Wireless Application Protocol (WAP) and how it supports various applications.
> 3. About security in m-commerce.
> 4. Payment methods in m-commerce.

8.1 INTRODUCTION

So far we have described e-commerce using desktop PCs and servers connected to the Internet. As we saw, the availability of e-commerce sites 24 × 7 allows anytime shopping from anywhere in the world. The desktop machines are connected to a LAN and are not portable. Nowadays people on the move want to use e-commerce facilities when they travel. There are two situations which arise. One is the use of a mobile laptop computer and the other is the use of mobile hand-held devices such as high-end mobile phones, or mobile Personal Digital Assistants (PDAs). A mobile laptop is used normally when a person is stationary, e.g., while waiting in an airport lounge for a plane or while working in a hotel room. In these cases a system called Wi-Fi (Wireless High Fidelity) connection is used to connect a laptop wirelessly to a wireless hotspot which is in turn connected to an ISP. Many airports, hotels and even city streets are "Wi-Fi enabled". In other words, they have wireless access points with a transceiver called *wireless hotspots* connected to the Internet through an ISP. The laptops will also have transceivers so they can connect using the hotspot to the Internet. (We have described these in Chapter 2). In this case the only additional problem requiring attention in e-commerce is the security of wireless connection. Otherwise, there are no new problems.

The number of mobile hand-held devices such as mobile phones is much higher than laptops. In India, the number of mobile phones in use far exceeds the number of desktops. It is estimated that there are about 400 million mobile subscribers in India as of September 2009 and are growing at approximately 8 million per month. There are several unique applications of e-commerce when mobile devices are used which we will describe in this chapter. Mobile phones use cellular wireless infrastructure to communicate with one another. The wireless cellular infrastructure is connected to the Internet and this allows e-commerce. Currently B2C e-commerce using sophisticated mobile hand-held devices such as PDAs is rapidly increasing. It also presents several new problems. They are:

- The screen size of mobile devices is very small. Thus, designing appropriate browsers is a challenge.
- The keyboard in mobile phones has only around 16 keys and thus several key strokes are required to send a message.
- The time to transmit messages using a cellular network is much higher (>200 ms) and is variable compared to fixed networks.
- The computing power of hand-held devices and available memory is much less than desktops/laptops. This requires innovative programming.
- Wireless systems can be easily intercepted compared to a wired system. Security requires special attention.

Thus, the standards and method used in fixed networks are not directly applicable in mobile systems and new systems are needed.

In this chapter, we will first present a layered architecture to logically describe m-commerce system. We will then describe several new applications which emerge when mobile devices are used. We will lastly discuss the essentials of physical, logical and mobile services layers.

8.2 LAYERED ARCHITECTURE FOR M-COMMERCE

In Chapter 2, we described a layered architecture for e-commerce. The reason we use a layered approach to describe e-commerce is the realization that each layer provides a service which is reasonably self-contained. However, the higher layers depend upon the services provided by the lower layers. Further, each layer can be designed independently assuming that services offered by the lower layers are available. In m-commerce, the major difference is the use of the mobile telephone infrastructure which permits mobility. Once the mobile device gets connected to the Internet, the services of the Internet which were used in e-commerce are available. Thus, we have to have layers which are appropriate. We propose a layered architecture which is adapted from the one proposed by Varshney et al. (See Reference 21 at end of the book). In this model there are four layers shown in Table 8.1.

Table 8.1 A layered architecture of m-commerce system

Layer name	Services provided
Application layer	• B2C m-commerce • Location based services • Logistics • m-payment system
Services for mobile hand-held devices	• Wireless Markup Language • Browsers
Logical layer	• WAP gateway services • Wireless Application Protocol stack
Physical layer	• Wifi based LAN • Cellular wireless system • Satellite based system (All the above are called Bearers)

We will discuss in the rest of this section various applications.
We broadly classify the applications as those using:

1. Mobile phone infrastructure and primarily based on Short Messaging Service (SMS) available on all mobile phones.
2. Those which use mobile laptop computers using Wi-Fi connection to LAN. There are portable light weight mobile laptops which are available now with 7-inch screen and pen drives which are useful for mobile applications.
3. Mobile high-end devices which are WAP-enabled using cellular network.
4. Global Positioning System (GPS) along with WAP-enabled mobile phones.

Each of these has services which are unique and are in various states of maturity. We will now describe applications in each of these categories.

8.2.1 Mobile Phone—SMS System

This is quite mature and is used extensively. Applications include downloading ring tones by sending an SMS, SMS confirmation of airline ticket booking, SMS alerts on delayed flights, SMS alerts on traffic jams, SMS messages on cricket scores and many similar services offered by mobile phone operators. The main limitation is 160 characters per message. Charges are debited from prepaid SIM cards in the mobile phone or billed if it is a post-paid service.

8.2.2 Laptops using Wi-Fi LAN Systems

As laptops accessing the Internet have large screens and reasonable computing power, they can be used for several applications described in Chapter 1. One of the major applications of this mobile system is that it enables a businessman while he or she is travelling to download e-mail, obtain information on latest inventory levels before committing a delivery deadline, get access to catalogues and price lists and generally to keep in touch with his or her reporting office. Wi-Fi security is however a problem.

Another application of a mobile laptop is in hospitals which are Wi-Fi enabled. A doctor treating a patient can retrieve the medical data of the patient from a patient's database. He or she may use it to prescribe appropriate medications which can be entered using the laptop. This will update the patient's database and keep the medical history of a patient up to date.

8.2.3 WAP-Enabled Mobile Hand-held Systems

Nowadays one can purchase mobile phones which are useful to send digital data besides being used for normal voice services. These phones have software which is incorporated in them called Wireless Application Protocol (WAP). WAP is the analog of TCP/IP protocol stack of the Internet. Thus, WAP enabled phones, which have more CPU power and memory compared to voice phones, may be used to create a mobile Internet. With the mobile Internet m-commerce can be realized in the mobile domain. This is not as useful as being able to use traditional Internet-based web services which have matured over years. Thus, the mobile Internet is connected to the Internet by mobile network operators. We will describe how this is done later in this chapter. We are, however, assuming that this connection exists when we describe applications.

There are several novel applications of WAP-enabled mobile hand-held systems using the cellular infrastructure besides the normal B2C e-commerce. In B2C e-commerce a customer can use the browser on the hand-held device to log on to the web sites of several stores in the vicinity where he or she is currently located and find out about availability and cost of items he or she wants to buy. A customer can participate in auctions using his or her mobile device and bid for goods while on the move.

Another useful application is mobile banking. Customers may log on to their account statements maintained by the bank (with a password in addition to the mobile phone number) and download their account details. Standing instructions such as debiting the account to pay instalments for a purchase or credit card payment or stop payment of a cheque may also be issued by a customer from the mobile device. On-line trading in shares while on the move is another popular application.

Salespersons can use their mobile devices to find out about items available in stock, negotiate discounts and record the sales with their company to enable the company to fulfill the order expeditiously.

A service engineer while repairing a machine can log on to the company to get a trouble-shooting manual on-line using his or her mobile device and may also get expert advice if there is a difficult problem.

Mobility provides access from *anywhere while on move* in the coverage area of the cellular infrastructure. This unique advantage will lead to many innovative applications not described in this section.

8.2.4 Location Dependent Services

A mobile unit's location may be approximately found by using the location of the Base Station of the cell which relays signals to it. When this Base Station's location is overlaid on a map of a city, the approximate position of the mobile device in a city can be found. This can be used to find, e.g., restaurants in the vicinity of the mobile phone with their menus and price and customer feedback (if available in the restaurant's web site) to select an appropriate restaurant. A shopper may find availability of goods he or she wants to purchase by logging on to the web sites of shops in the vicinity from the map, compare prices and visit the selected shop.

There are several other applications which are location dependant and require more accurate location information. In these cases a mobile device may be enabled with a built-in receiver of Global Positioning System (GPS) and find the exact longitude and latitude of the device. By overlaying this on a map, the exact location of a mobile device may be found. These applications are:

- An empty taxi in a fleet may send location information to a dispatch centre using a mobile hand-held device. The dispatch centre can then send the nearest empty taxi to a customer.
- A request may come to a company to ship urgently spare parts for a machine. The company, knowing the location and contents of a moving truck with the required spare parts can send a message to the truck driver to deliver the items to the customer.

8.3 MOBILE COMMUNICATION INFRASTRUCTURE

While describing the layered architecture we saw that the bottommost layer which is called the physical layer provides the infrastructure for mobile communication. Among these we have described Wi-Fi based LAN and satellite-based system in Chapter 2. In this section, we will very briefly describe the mobile phone infrastructure which supports mobile hand-held devices. The primary objective of a cellular wireless communication system is to allow customers to use their own personal mobile devices while they are on the move. The handset is small enough to go in their pocket and a customer will be connected to a communication system anywhere any time (if he or she chooses to). Even though the cellular communication system was intended for voice communication it is currently a digital system and is capable of digital communication. Handsets have also evolved which can compute as well as communicate. There are two cellular communication technologies which are both popular and coexist. A system called Global System for Mobile Communication (GSM) is used in most of Europe, far East and in India by BSNL, Airtel, Spice, etc. The other system called Code Division Multiple Access (CDMA) is popular in USA, Japan and used in India by Reliance Mobile and Tata Indicom. Both GSM and CDMA use cellular wireless technology but the modulation methods are different. As the modulation systems are different, the hand-held devices use different technologies and they are not interchangeable. In this section, we will first explain the overall architecture of a cellular communication system and how mobility is ensured by using inter-meshing hexagonal cells to provide wireless coverage for whole regions.

8.3.1 Architecture of GSM Cellular Mobile Wireless System

In GSM mobile systems the region to be covered by the service is divided into hexagonal cells which are typically 5 km in diameter (See Figure 8.1). The size depends on the density of traffic and would be larger in sparsely populated areas. At the centre of each hexagonal cell is a base transceiver station (BS) which sends and receives signals to/from mobile devices within its range. The cell shape is hexagonal as the distance between the centre of neighbouring hexagons is constant. This property is useful to ensure easy handoff, i.e., transfer of control of a mobile device (phone, PDA, etc.) from one cell to an appropriate adjacent cell when the user of the mobile device moves. The base station has several antennas, a controller and a set of transceivers, one transceiver for each channel assigned to that cell. The base stations are all connected to a Mobile Telecommunication Switching Office (MTSO) usually by landlines. If the terrain is unsuitable BS may be connected to MTSO by wireless. The MTSO is also connected to a PSTN by a landline (See Figure 8.2).

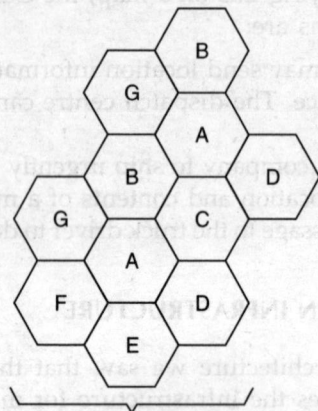

A group of cells A, B, C, D, E, F, G having distinct frequencies

Figure 8.1 Cellular structure (cells which are minimum two cells away use the same frequencies).

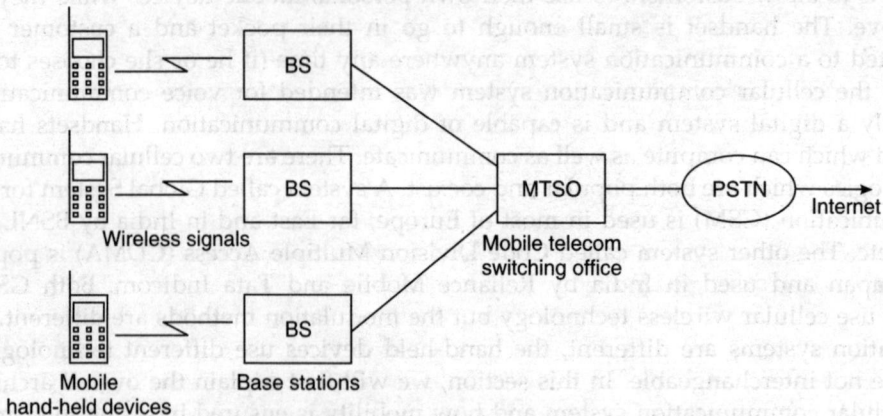

Figure 8.2 Architecture of mobile telecom system.

The major functions of MTSO are to:
 (i) Establish connection between a fixed phone connected to PSTN and mobile device via BS and vice versa.
 (ii) Establish connection between two mobile devices by using respective base stations.
 (iii) Allocate appropriate channels to mobile devices to communicate by informing concerned BS.
 (iv) Manage handoff of mobile devices from a BS to an appropriate one when a mobile user moves.
 (v) Monitor the calls in progress for facilitating charging for use.

Before describing how the connection is established between mobile devices and fixed phones as well as between two mobile devices, we will have to understand how channels are allocated to mobile devices when they want to communicate. The most expensive resource in the mobile system is the electromagnetic spectrum allocated for this purpose. A specific band of frequencies is allocated to a mobile network operator by the government which has to be used judiciously by the operator to support as many mobile subscribers as possible. In a GSM system, this band is divided into a set of sub-bands and one set is allocated to each cell for communication within that cell. As mobile devices have a limited power (due to their small size and need to conserve battery life) the cell sizes are small (few km) and their transmission has a limited range. Thus, a sub-band of frequencies allocated to a cell can be reused in other cells which are not in their immediate vicinity. This frequency reuse is a very important feature of the cellular communication system. We will now explain how a connection is established in a cellular system. We will first describe connection establishment between a fixed phone and mobile phone and vice versa. The following steps are followed:

Step 1: A request to contact a mobile device is received by MTSO and is relayed to all base stations (BSs) connected to it.

Step 2: The BSs send out a paging signal in their respective bands seeking an acknowledgement from the called mobile device.

Step 3: The called mobile unit recognizes its number in the cell in which it is currently located and sends an acknowledgement to the BS of the cell. The BS in turn sends this information to MTSO. MTSO establishes connection of this BS with the fixed phone via PSTN. It also allocates a free channel to the BS to communicate with the mobile device. Connection is established between a fixed phone and the mobile device via MTSO and BS.

Step 4: If the mobile device moves away from the cell whose BS established the connection, it picks up a new signal from the BS into whose domain it has shifted. This BS now allocates a channel to the mobile device and connects it to MTSO. This is called handoff from one BS to another.

The above procedure is for a call from fixed to a mobile device. If a mobile device wants to call a fixed phone, it establishes connection with the BS whose signal is strongest and connects to MTSO via this BS. MTSO assigns a channel after it finds that the fixed phone is free to receive a call. Now the call can proceed. Handoff is the same as in step 4 above.

The steps followed for calls from a mobile device to another mobile device are as follows:

Step 1: A mobile device wanting to call is turned on and senses wireless signals available on various frequencies and picks the one which is strongest. It then selects the BS sending this frequency as its base.

Step 2: The BS forwards the called mobile number to MTSO. MTSO relays this to all the base stations. The BSs in turn send paging signals in their respective bands seeking an acknowledgement from the called mobile number.

Step 3: If the called mobile number is ON, it picks up the paging signal and sends an acknowledgement to the BS in whose cell it is currently located.

Step 4: The BS sends this information to MTSO which connects the BSs of the two mobiles so that conversation or a digital transaction between them can proceed. If in step 3 no acknowledgement is received by any BS due to the called mobile being switched off or busy or out of range, this information is sent to MTSO which relays this information to the calling mobile via its BS.

Step 5: If any or both of the mobile units move from the current cell, a procedure similar to that described in step 4 of fixed to mobile unit is followed. All the above steps take place fast and once a connection is established, it is maintained regardless of movement of the mobile phone and without any obvious disruption of the ongoing conversation.

Cellular communications systems used nowadays, namely GSM and CDMA are both digital systems and thus all the steps described above are for digital data communications. The frequency spectrum used in cellular systems in India is in the 900 MHz band.

In the above description, we have described the cellular phone system which has been designed primarily for phone conversations. In m-commerce, the cellular system is used by mobile devices as the "bearer network" of signals to access the Internet. In this case there are two alternatives. One is for the mobile network operator to provide Internet services from MTSO and the other is to let another independent Internet service provider to connect via MTSO to a gateway which does protocol conversion of the mobile Internet to the Internet on landlines. We will discuss the various alternatives available with other types of "bearers".

8.3.2 General Packet Radio Service (GPRS)

A packet data-oriented mobile *data service* is available in GSM systems and is called GPRS. This is popular as the cost of a mobile device with added GPRS is in the medium price range (around Rs. 15000 in 2009) compared to higher cost of WAP-enabled mobile systems. GPRS conncectivity can also be provided on laptops enabling them to use the mobile telephone network. The transmission rate of GPRS is in the range of 56 to 114 Kbps (the speed is increasing). GPRS can be used with WAP but can also be used independently to avail of Internet services such as e-mail and access to World Wide Web. GPRS tariff is per MB of data sent/received and is around Rs. 3 per MB. This is to be contrasted with mobile phone use which is usually charged for connection time. For several applications GPRS is preferable. An improved GPRS called EDGE (Enhanced Data rate GSM Environment) is now being offered by some operators and delivers data at rates of up to 384 Kbps. The next generation mobile communication system called 3G is expected to give data at rates up to 2 Mbps for packet communication.

8.3.3 CDMA 1xEVDO Rev.A

CDMA 1xEVDO Rev.A (Evolution-data optimized) provides digital packet data communication for CDMA networks. It provides packet data communication for mobile subscribers with an average speed of 300 to 500 Kbps (peak 2.4 Mbps). Laptops can also use CDMA 1xEVDO connectivity and use the mobile network. With the emergence of 3G networks it is expected that the speeds will go up. The charging is not for connect time but for the amount of data traffic. The rates are around Rs. 4 per MB. This will be over some fixed price for the connection and various plans would be available from competing mobile phone providers.

Normally, GPRS and CDMA 1xEVDO are used by laptop computers or expensive mobile devices to access the Internet using cellular communication infrastructure. Low cost mobile phones do not use these. As laptops are powerful, VPN connectivity can be provided to them using GPRS or CDMA 1xEVDO.

8.3.4 Short Message Service (SMS)

This is a communication protocol which enables the interchange of short text messages between mobile phones/devices. It is an extremely popular service. Messages are sent to a Short Message Service Centre (SMSC) which stores it and forwards it to the intended recipient's phone. The service is best effort, i.e., delivery is not guaranteed but normally messages are not dropped. SMSCs store and forward the message to a recipient's phone if it is not reachable or is switched off. The maximum length of a message is 160, 7-bit character or 140 bytes. If unicode is used (e.g., to send non-English characters), then the message length is limited to 70 characters.

SMS gateways provide value-added services in cooperation with mobile network operators to businesses (such as TV voting). These operators route messages between different networks and provide faster SMS delivery.

M-commerce in India uses SMS widely as they are low cost, universally available and matches the available technology. Several innovative services can be offered to a clientele which does not have Internet access.

8.4 WIRELESS APPLICATION PROTOCOL

The Internet was enabled by the adoption of a common standard called TCP/IP (Internet Protocol Stack) by manufacturers of all computers. In the early days, a mobile communication system was mainly intended and designed for audio phone conversations. When mobile devices became more sophisticated and the networks became all digital, it was realized that digital data-oriented services could be offered. All the applications, we described in Section 8.2 require the mobile hand-held device to avail the services of the Internet. In order for the mobile devices of different manufacturers connected by the wireless network to communicate with one another seamlessly, we need a protocol similar to TCP/IP. TCP/IP cannot be adopted without change due to several reasons, the most important one being the high and variable time (>500 ms) needed to transmit messages among mobile devices and from mobile devices to the fixed network. Packet losses are also higher in wireless networks. Thus, another protocol called *Wireless Application Protocol* (WAP) has been adopted by manufacturers of mobile devices. With the adoption of this standard, we have a "Wireless Internet" in the wireless world to communicate among wireless devices. For wireless devices to communicate with services connected to the Internet, we need a protocol conversion from WAP to TCP/IP and vice versa. This conversion is performed by a device called *WAP gateway*. Before describing the functions of a WAP gateway, we will briefly describe the WAP protocol stack.

WAP stack is also a layered protocol similar to TCP/IP of the Internet. The advantages of layering are:

1. Layering allows the design of each layer independently of the other layers.
2. Layering allows subsets of the standard to be implemented by equipment manufacturers.

3. Layering permits easy bridging to the Internet.
4. Layering separates concerns of functions to be provided by each layer.

8.4.1 Mobile Network Operators

There are several mobile network operators in each country. The technology used also differs. We pointed out that GSM and CDMA are two different systems. Their original purpose was as voice carriers. Now they are evolving as data carriers as there are many more value-added services they can provide.

8.4.2 Mobile Handset Manufacturers

There are several manufacturers each making several models. Low-end models have only around 16 keys and a small 3.5 cm × 2.5 cm screen. High-end devices have over 36 keys and a 8 cm × 6 cm screen. They also manufacture models compatible with different network standards.

8.4.3 Service Provider

Service providers provide several application services to mobile users. They cooperate with content providers, e.g., banks, e-shop, etc. Several services have already attained maturity in the fixed client-based Internet. They have to adapt them to mobile clients who have handsets which have much less capacity. The effort will be worthwhile only if the subscriber base is large.

The Wireless Application Protocol has been designed primarily cooperatively between network operators and mobile handset manufacturers. The layering idea allows each stakeholder to independently design layers. The design would be optimized for their specific device technology. However, as the interfaces are clearly defined, they can mesh and depend on services provided by the lower layers. The services provided by each layer of WAP is summarized in Table 8.2. In Table 8.3, we compare WAP with Internet Protocol. Observe the close correspondence between the two protocols.

Table 8.2 WAP protocol stack

Layer Name	Services
Application Layer (WAE)	• Wireless Markup Language (WML)
Session Layer Wireless Session Protocol (WSP)	• Supports efficient long-term session between client and WAP gateway • Allows efficient operation of WAP micro browser over low bandwidth, high latency network
Transaction Layer Wireless Transaction Protocol (WTS)	• Explicitly pairs client request and server response • Reliable message exchange • Optimizes bandwidth use
Security Layer Wireless Transport Layer Security (WTLS)	• Similar of TLS/SSL of IP • Public key Encryption • Defends against security attacks such as replay attacks and denial of service attacks

(Contd.)...

Table 8.2 WAP protocol stack *(Contd.)...*

Layer Name	Services
Datagram Transport Layer Wireless Datagram Protocol (WDP)	• Hides differences between underlying bearer networks • Sends datagrams point to point • Similar to user datagram protocol • Bearer network dependent • Does not guarantee ordering, reliability
Physical Layer (Bearers)	• Mobile Telephone Services GSM, CDMA • Mobile Data Services GPRS, SMS, CDMA 1xEVDO

Table 8.3 Comparison of WAP protocol stack and Internet Protocol stack layers

Layer of WAP	Layer of IP	
Application/Browser (WAE)	Application/Browser	
Session (WSP) Transaction (WTP)	Transaction/Session (HTTP)	
Security (WTLS)	Security (SSL/TLS)	
Datagram Transport (WDP)	Connection oriented (TCP)	Datagram transport (UDP)
Bearers	Media	

Looking at the application layer the main component is WML which is defined as XML 1.0 application. As the size of the screen in most mobile hand-held devices is small the design of WML should recognize this. The microbrowser also is for a small screen. The most apt applications for hand-held mobile devices are simple ones requiring a small output which can be accommodated in a small screen such as train/airline departure/arrival information, looking at stock prices, looking up addresses and phone numbers of shops in the vicinity of a user, traffic information, etc.

WAP was originally developed as a cooperative effort of mobile network operators and mobile phone vendors. A group called WAP Forum did this. Now WAP Forum has been succeeded by a company called Open Mobile Alliance (OMA) which will continue to improve WAP as new networks (3G, for example) emerge and better hand-held devices are built.

8.5 WAP GATEWAY

We saw in the last section that the mobile hand-held devices use WAP protocol to communicate with one another seamlessly regardless of the technology used by mobile network providers (CDMA or GSM). In other words, WAP is the analogue of TCP/IP of the fixed networks. If all services were available on mobile networks, it would be a self-contained system. A mobile system cannot however work in isolation. It has to communicate with the Internet with fixed clients and servers. WAP-enabled mobile clients would require services offered by servers connected to the Internet. In order to achieve this, the WAP transaction/session has to map to IP transaction/session which uses http. This is done by WAP gateway which transforms WSP requests received via the wireless network into http request and sends it over the Internet. In the reverse direction, it transforms http responses received from the Internet to WSP responses to the requesting WAP device via the wireless network (See Figure 8.3).

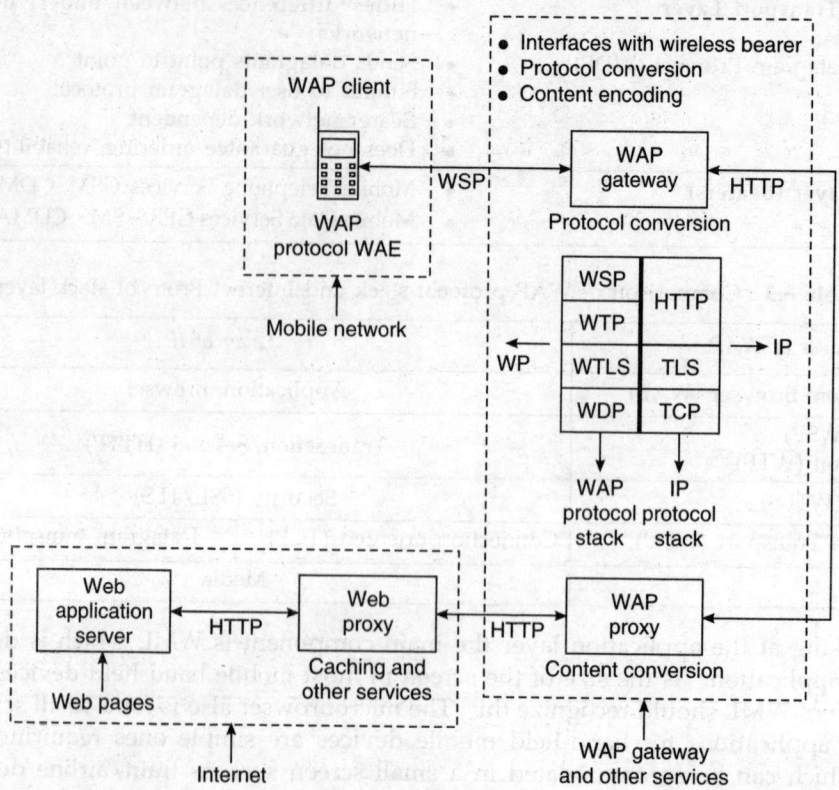

Figure 8.3 WAP gateway and proxy connecting WAP client and web application server.

Mobile hand-held WAP-enabled devices cannot handle HTML which is the most common language used to write pages in the World Wide Web. As the screen size is small, memory capacity is low and CPU power is limited only 3 or 4 lines can be displayed on a hand-held mobile device. The language used for display is called Wireless Markup Language (WML) which is defined as an XML 1.0 application. Besides WAP gateway, a WAP proxy server is required to connect and present HTML pages from a web site to WML formats. (Some service providers present applications needed by mobile users as WML pages or decks as they are called. This expedites access to information to mobile devices). The WAP gateway recognizes the fact that wireless networks have low bandwidth and high latency. It thus encodes the information optimally to cater to this constraint.

A mobile device supports a microbrowser which is a software that requires a small memory, low computing power and a small screen size. They use WML. Thus, WAP proxy must act as a WML server to a mobile client using the information retrieved from the specified URL as HTML pages.

Each MTSO in the physical system would usually have a WAP gateway and WAP proxy to cater to Internet service request from WAP-enabled mobile devices in its network.

8.5.1 WAP and i-Mode

WAP is a standard developed by WAP Forum and now transferred to Open Mobile Alliance Ltd. It is popular in USA, Europe and India. In Japan the standard used was developed by NTTDoCoMo. As we have seen WAP is an open, global specification. It uses WAP as a bridge between the mobile network and the fixed network.

i-mode, on the other hand, is a packet switched mobile Internet service for mobile devices made by NTTDoCoMo of Japan. In Table 8.4, we compare WAP and i-mode.

Table 8.4 Comparison of WAP and i-mode

	WAP	i-mode
Developer	Open Mobile Alliance	NTT DoCoMo
Status	Open standard	Proprietary
Function	A protocol	A full mobile Internet Service
Language	WML/XHTML Wireless Markup Language	CHTML (Compact HTML)
Access to fixed Internet	WAP Gateway and WAP Proxy	TCP/IP modifications

8.6 WIRELESS MARKUP LANGUAGE

We saw that mobile clients use a Wireless Markup Language for displaying information on their screens. WML has inherited many features of HTML but is based on XML and thus its syntax is much stricter.

WML is used to create pages that can be displayed in WAP browser. Pages in WML are called *decks*. A set of *cards* make up a deck. Cards are related to one another by links. When a WML page is accessed by a mobile hand-held device, all the cards in the page are downloaded into its memory from the WAP gateway server. Navigation between cards in the page can be done locally without using the wireless network. We give in Figure 8.4 an example of a WML program called a document.

```
<? xml version ="1.0"?>
<! DOCTYPE wml PUBLIC" -//WAP.FORUM//DTD.WML 1.1//EN"
http://www.wapforum.org/DTD/wml_1.1.xml>
<wml>
<card id = "WML" title = "What is WML?">
<p>
This section answers the question:What is WML?
</p>
</card>
<card id = "Syntax" title = "WML Syntax Rules">
<p>
Some of the essential WML Syntax rules are explained
</p>
</card>
</wml>
```

Figure 8.4 A simple WML document.

We explain this document in what follows. The prologue to the document is given below:

```
<? xml version ="1.0"?>
<! DOCTYPE wml PUBLIC" -//WAP.FORUM//DTD.WML 1.1//EN"
http://www.wapforum.org/DTD/wml_1.1.xml>
```

The first line states that this is an xml document and gives the version. It is followed by the address of the DTD to be referenced. The part enclosed by the tags <wml> </wml> is called a deck. It encloses the entire WML document. Each card in the deck is enclosed by <card> </card> tags. The paras are inside <p> and </p> tags. Each card has an id and a title. On the mobile it will be displayed as shown in Figure 8.5. Observe that each card is one screen display. We have shown the message in two lines is the screen. It may, however, be displayed on several lines depending on the mobile device's screen size.

```
—--What is WML?
This section answers the question: What is WML?
```

Screen 1

```
---WML Syntax Rules —
Some of the essential WML syntax rules are explained
```

Screen 2

Figure 8.5 Display of 2 cards on mobile display.

WML has several other tags similar to HTML and the document formatting structure is similar. We give a few examples below:

```
<wml>
    <card title = "Paragraph">
    <p>
    This is a paragraph <br/> with a line break
    </p>
    </card>
</wml>
```

The output will look like:

```
- - - Paragraph - - - -
This is a paragraph
with a line break
```

Figure 8.6 Display of paragraph with a break.

It is important to observe that we have used lower case letters. WML is case sensitive. WML has features to represent a table with tags. The tags are similar to HTML tags

<table>, <tr>, <td> for table, table row and table data. It also has the anchor tag <a> to link to another page to enable non-linear reading. It also has the image tag <img src = "pic.wbmp"/> to display an image. Observe that the only image type that can be displayed in .wbmp. WML has also a feature to accept information from the user of a mobile device. A simple example is illustrated in Figure 8.7. Observe how input data is restricted to be numeric for Roll No.

```
<wml>
<card title = "INPUT">
<p>
Roll No: <input name = "Roll No" size = "15" format = "*N"/><br/>
Name : <input name = "Name" size = "15"/><br/>
Course : <input name = "Course" size = "15"/>
</p>
</card>
</wml>
```

Figure 8.7 WML document to input data.

The result on a display may look like that shown in Figure 8.8. 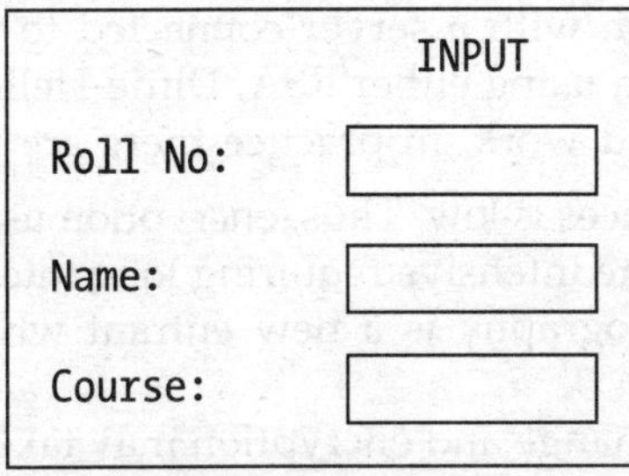

```
         INPUT
Roll No: [      ]
Name:    [      ]
Course:  [      ]
```

Figure 8.8 Display for WML document of Figure 8.7.

It also has features to select one or more menu items from a displayed list. There are tags to represent actions such as jump to a new card with <go> and go back to previous card with <prev> tag.

As the purpose of this section is to illustrate some of the features of WML and not as a tutorial to enable you to write WML code we have left out a lot of features such as fonts, activating tasks using <do>, timer, setting a variable, etc. Those interested in a quick tutorial may go to the site **www.wapforum.org**.

8.6.1 XHTML

XHTML is a new markup language which is a combination of HTML and XML. It has all the features of HTML 4.01 combined with the syntax of XML. HTML has loose syntax rules and is forgiving when minor errors are made whereas XML is strict. For resource constrained hand-held mobile devices XHTML is implemented by some vendors with mobiles with larger screen sizes. The most important differences between HTML and XHTML are:

- XHTML elements are properly nested, e.g, <li> and must have </li> for listing elements.
- XHTML elements must always have a closing tag for all elements.
- XHTML is case sensitive. Lowercase is used for all elements.
- It must have one root element <html>. The program must be enclosed by <html> …</html>
- <!DOCTYPE> is mandatory

We will not describe XHTML in detail. Those interested must look up World Wide Web consortium web site (www.w3.org/TR/html2)

8.7 SECURE WIRELESS CONNECTIVITY

Wireless security requirements are the same as in wired network. We require authentication of who is calling, protect the data from eavesdroppers while it is in transit and ensure that it is not altered while it is transmitted. There are some unique problems in wireless: eavesdropping is easy, bandwidth is limited, latency is high and connections are unstable. In WAP devices security is provided by Wireless Transport Level Security (WTLS) which works in a manner similar to SSL (Secure Sockets Layer) of Internet Protocol. The simplest authentication is server authentication in which a server provides to a client a public key certificate. Once a client is assured that the server is genuine, the client can use a name/password based transaction with a server connected to it by the mobile network. Even though in theory, encryption using either RSA, Diffie-Hellman session key exchange or RSA/3DES combination should work, in practice there are the following problems:

- CPU power of mobile devices is low. Thus, encryption using RSA or Diffie-Hellman is slow as they are compute intensive requiring long integers to be raised to a large power. Elliptic key cryptography is a new entrant which is expected to be less compute intensive.
- Network is slow. Key exchange and encryption may take several seconds. With 3G networks this would be reduced.
- WTLS is applicable within the mobile network only. If a mobile client has to access a server on the Internet, WTLS has to be converted to TLS at the WAP gateway and new problems arise which we will discuss next.
- In WAP protocol stack WTLS is optional and need not be used if not considered essential.

8.7.1 Security of Mobile Network-Internet Connection

WTLS applies only within the mobile network. If a mobile client wants to connect to a server on the Internet, protocol conversion from WAP to IP is performed by the WAP gateway. In order to convert data encrypted by WTLS to SSL/TLS encryption, WAP gateway has to first obtain the plain text of the WTLS encrypted data and re-encrypt it in SSL/TLS. Thus, there is no end to end security between a mobile WAP client and a server connected to the Internet. Normally, the plain text is available only for a few milliseconds and that too only in the main memory of the WAP gateway (WAP gateway will never put the plain text on disk). Thus, a hacker has to get the privileges of a super user to enter main memory. Further, the WAP gateway normally protects itself with a good firewall which makes it difficult to access it. Thus, this problem should not unduly perturb a user. However, if sensitive financial transactions are involved there are other methods of connecting mobile client to the Internet which we discuss next.

8.7.2 WAP Gateway Managed by Sensitive Content Providers

In this case a service provider can cooperate with the mobile network operator to connect another WAP gateway entirely under its control ensuring security. Payment gateways of some banks which provide mobile banking (See Figure 8.9) use this solution.

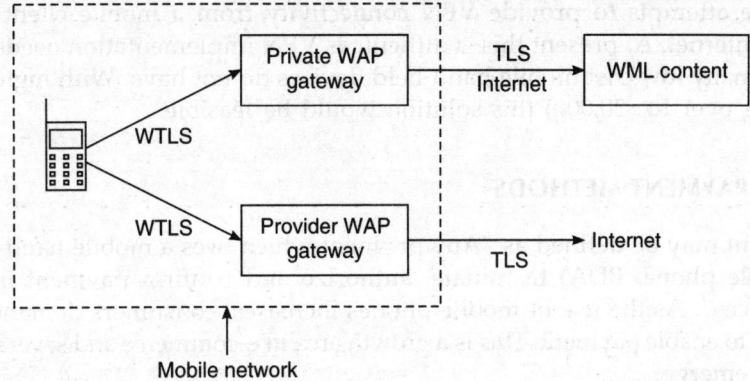

Figure 8.9 Private WAP gateway.

8.7.3 WAP Gateway at Server End

Another method is to use a "tunnelling gateway" at the mobile network rather than a WAP gateway. This tunnelling gateway takes the full WAP session, transport and security protocols in the WAP stack of the mobile network and encapsulates them as WDP packets. WDP packets are transmitted as SSL encrypted UDP packets on the Internet (See Figure 8.10) to the server and a WAP gateway converts WAP protocol to IP and processes the mobile client's request. This solution ensures end to end security of WAP packets.

This solution, however, places a huge burden on the server as it has to maintain WAP gateway software and also provide contents in wml.

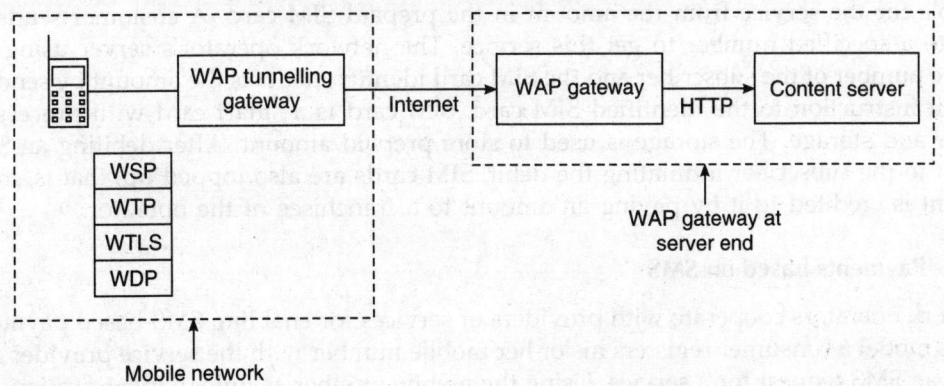

Figure 8.10 Tunnelling WAP protocol stack to a remote WAP gateway at the server.

To ensure end to end security between a mobile client and a server and for mutual authentication, public key certificates should be exchanged between the two parties. The certificates are too long and transmission is slow. In the case of content provider maintained WAP gateway, the authentication may be done only once and then RSA/3DES encryption can be used. In other cases the mobile client can store its certificate in a web site on the Internet and send its URL to the server being accessed. To some extent this alleviates the problem. However, in most cases the users are casual users (usually in B2C e-commerce) and do not have a public key certificate. In such cases Diffie–Hellman session key exchange (See Section 5.3) and 3DES encryption of the messages is used.

There are attempts to provide VPN connectivity from a mobile client to the web server on the Internet. At present this is difficult as VPN implementation needs computing power which many low cost mobile hand-held devices do not have. With high end mobile phone (costing over Rs. 30,000) this solution would be feasible.

8.8 MOBILE PAYMENT METHODS

Mobile payment may be defined as "Any payment which uses a mobile hand-held device (such as mobile phone, PDA) to initiate, authorize and confirm payment in return for goods or services". As the use of mobile phones increases, consumers demand the use of mobile phones to enable payment. This is a growth area in e-commerce and several innovative solutions will emerge.

All mobile payment systems must have the same properties as e-payment system (See Section 6.2). They are confidentiality, integrity, authenticity and non-repudiability.

There are several types of mobile payment systems. They are:

- SIM (Subscriber Identity Module) card enabled
- Payment using Short Messaging Services
- Payment using WAP enabled mobile hand-held device
- Using Smart card (See Section 6.6.3) along with the mobile hand-held device

8.8.1 SIM Card-enabled Payments

SIM is widely used in India for small payments. Mobile network operators sell value-added services such as ring tones and access to cricket scores by deducting the amount payable for the service from the amount in the prepaid SIM card. A customer sends an SMS to a specified number to get this service. The network operator's server using the mobile number of the subscriber and the SIM card identity, deducts the amount by sending a debit instruction to the identified SIM card. SIM card is a smart card with processing power and storage. The storage is used to store prepaid amount. After debiting an SMS is sent to the subscriber intimating the debit. SIM cards are also topped up, that is, more amount is credited to it by paying an amount to a franchisee of the operator.

8.8.2 Payments based on SMS

Network operators cooperate with providers of services for enabling SMS-based payment. In this model a consumer registers his or her mobile number with the service provider and sends an SMS request for a service. Using the mobile number as unique identification, the service provider bills the customer after providing the service. In case of dispute authentication of the mobile number is provided by the network operator. Normally, these types of SMS-based payments are micropayments, namely, payments below Rs. 1500.00. SMS security depends on encryption provided by the Mobile Network Operator. This is good enough for small payments. The main disadvantage of SMS is that the message length must be below 160 characters.

As most mobile phone users in India do not have expensive phones and many of them would like to use their mobile phones for shopping, some innovative methods are emerging and are now growing. Some merchants find it attractive as transaction charges with mobile phones are normally lower than with credit cards. However, Reserve Bank of India (RBI) has recently issued guidelines which restrict mobile phone payments to Rs. 10,000.00 per day primarily due to security concerns. Also RBI has mandated that the service should not be network operator specific. In other words, the service must treat all mobile operators equally.

Unfortunately, currently there are no accepted standards for this type of payment. One model for mobile phone payments for goods purchased from a merchant would have the following steps.

Step 1: A merchant accepting mobile phone payments has to register with a payment service provider giving his or her bank details and will be assigned a unique ID and his or her mobile phone number will be registered for payment confirmation. Similarly, an individual wanting to use mobile payment has to register his or her mobile phone number and bank account details with the same service provider. The service provider will assign a mobile personal identification number (mpin) to the customer.

Step 2: The merchant requests a customer wishing to pay using a mobile phone for his or her number and sends an SMS to him or her giving:

<merchant id><Transaction id><amount>

Step 3: The customer appends to this SMS his or her mpin and forwards it to the payment service provider authorizing payment.

Step 4: The payment service provider forwards it to the customer's bank for authorization. The bank will authorize and give authorization ID if this amount is within limits. It will debit customer's account.

Step 5: On receipt of authorization from the bank, the payment service provider will credit merchant's bank and forward SMS to *both* the merchant and the customer with following details:

<merchant id><Transaction id><amount><payment confirmation id>

Step 6: Merchant will now deliver the goods to the customer.

The security of the entire transactions depends on the encryption of SMS done by mobile operators which is reasonably secure for small payments. Dispute resolutions will be based on transaction details given in the last SMS of the payment service provider.

8.8.3 Payment using WAP-enabled Mobile Hand-held Device

We will describe credit card payment using WTLS (Wireless Transport Layer Security) which is the analogue of Transport Layer Security (TLS) of Internet (See Section 6.3.1).

There are several parties involved in enabling credit card payment. They are:
1. Customer using mobile hand-held device
2. The mobile network operator
3. WAP gateway and proxy which is at the edge of the mobile network
4. Land line (Internet) connecting WAP gateway to the merchant's server
5. Acquirer's server connected to the merchant's server by the Internet
6. Customer's bank (card issuer) connected to the acquirer by Internet

In Figure 8.11 we give a block diagram of the credit card payment system in m-commerce.

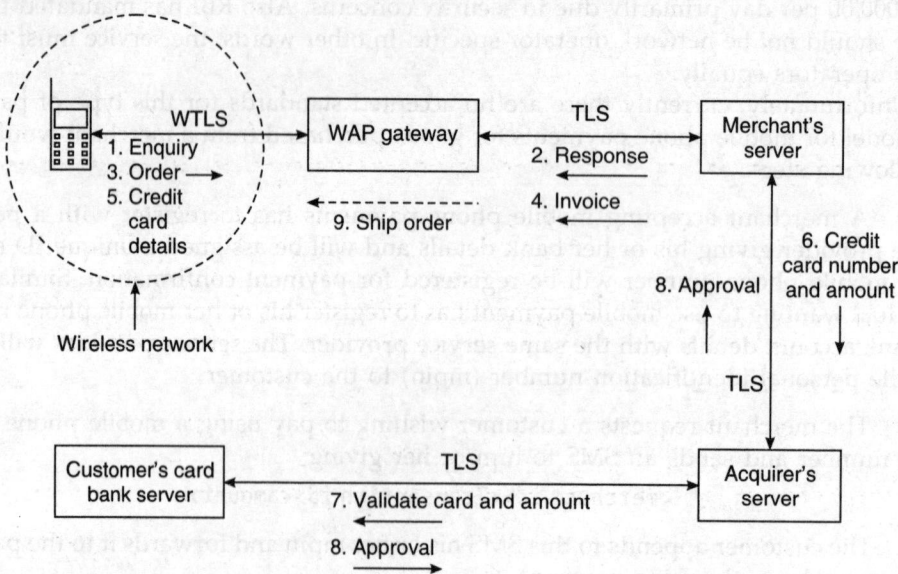

Figure 8.11 Credit card payment using a mobile hand-held device.

The payment method proceeds as follows:

Step 1: A mobile client accesses the web site of a merchant using his mobile device in which he enters the URL of the merchant.

Step 2: The merchant's website can be viewed using the mobile user's micro browser. The main problem is the small size of mobile screen. Thus, viewing can be painfully slow. If the merchant's server provides content using WML optimized for viewing by mobile user, it will be advantageous to the client. A few content providers cooperate with Network operators and provide such services.

Step 3: The mobile client now places selected items in his "shopping cart" with merchant and orders these.

Step 4: The merchant's server now prepares the invoice and sends it to the mobile client's device.

Step 5: The mobile client views it using WML. If he approves, he OKs the transaction. Now he has to send his payment. Normally, it will be by credit card. Credit card number has to be sent only on a secure channel. In the wired world TLS provides the necessary security. In WAP enabled hand-held device the equivalent is WTLS which we described in the last section. WTLS uses either RSA/3DES encryption or Diffie Hellman secure key exchange algorithm for security. Thus, the credit card details are sent securely to WAP gateways using WTLS. From WAP gateway to the merchant's server TLS security will be used. Along with encrypted credit card number the shipping address is also encrypted and sent to the merchant's server.

Step 6: The merchant's server decrypts the data. The credit card number and the amount are sent encrypted using TLS to the acquirer's server.

Step 7: The acquirer forwards the credit card details and amount to the appropriate customer's bank server using TLS.

Step 8: The customer's bank server validates card number and amount of order and informs the acquirer's server which forwards the approval to the merchant.

Step 9: The merchant ships the goods ordered to the customer. The customer's credit card company mails the bill for payment to the customer. After the receipt of the bill in due course of time the customer remits the amount to his bank.

In summary, the credit card payment method is very similar to the system used when an order is placed from a desktop client given in Section 6.3.1 where security was ensured by using Diffie–Hellman session key exchange for secure transmission. The only difference is the transaction between the mobile client and the merchant which uses WTLS instead of TLS. Even this security is difficult to implement unless the mobile client has reasonable CPU power to implement the Diffie–Hellman session key calculation algorithm. Due to the low bandwidth of wireless network coupled with low CPU power of mobile devices, the entire payment process is quite slow (tens of seconds). All transactions between the merchant, acquirer and the card issuing bank use TLS as they are all connected by the Internet.

SET protocol for credit card transactions presented in Section 6.3.2 is practically difficult to implement when a customer uses a mobile hand-held device.

In a payment system involving merchants and customers mutual authentication of the two parties is desirable. This is often provided by third parties maintaining trust/validation directories which can be referred to by the two parties before commencing transaction. These directories must be in encrypted form to avoid hacking and allow access only to approved clients using a password system.

8.9 MOBILE BANKING

Several private sector banks in India have started offering information based services like balance enquiry, record of last five transactions, instruction to stop payment of cheques, location of nearest branch/ATM, etc., for customers using mobile hand-held devices. They are also starting funds transfer instructions such as bill payment, transfer to other accounts, etc., provided customers pre-register their mobile numbers with the bank. Each customer is assigned an individual mobile personal identification number (mpin). Reserve Bank of India has issued operative guidelines for banks on starting mobile phone-based services (www.rbi.org). We will now give the gist of the method to be used by banks for carrying out banking transactions such as debit/credit to customer's account on the basis of funds transfer instructions received from hand-held mobile devices.

We will define mobile banking as: *"Any banking transaction such as funds transfer which is initiated by a customer using hand-held mobile devices and wireless telecommunication infrastructure"*.

In mobile banking, besides a customer, there are normally three more parties. They are the bank, mobile network operators and mobile payment service providers (also known as mobile payment gateway operators). Mobile network operators provide wireless infrastructure whereas mobile payment gateways provide technical support services to banks. Banks guarantee settlement of funds and ensure compliance with rules enforced by regulators in a country. Reserve Bank of India is the regulator in India. By definition of mobile banking, we have excluded services which only use mobile phones with attached devices such as smart cards and Radio Frequency Identification Devices (RFID). The reason is that they do not present any new technical problems compared to payment devices using the Internet. The security issues when using mobile devices which employ wireless transmission and mobile network operators are unique.

As we pointed out in the beginning of this section, SMS-based services from registered mobile phones are normally restricted to providing information and do not entail any special security problem. They are not secure for money transactions. However, for small payments (less than Rs. 5000 per day as specified by Reserve Bank of India) SMS is allowed. For transactions above this amount special security precautions are to be taken. We depict in Figure 8.12 the parties involved in the following type of services in which high end WAP-enabled mobile phones are used.

1. Debit an account and credit another account based on instructions by a customer.
2. Stop payment instruction for large amounts.
3. Any other standing instructions affecting the balance in one's account.

In WAP-based services, the customer must first give an account number and password registered with the bank through its proxy, namely, mobile payment gateway service operator appointed by a bank. Besides this to access any service a separate mobile personal identification number (mpin) has to be registered. This is a second factor in authenticating a customer. Normally, a table of mpins and corresponding account numbers must be stored in a hardware module (a Read Only Memory which cannot be hacked) by the Payment Gateway's server. Observe that several banks may operate mobile payment services and each bank may have its own proxy gateway. For inter-bank funds transfer the banks must have an arrangement so that these gateway operators cooperate or they may identify a single clearing service (just like RBI for cheques) to handle inter-bank payments and reconciliation. It would be ideal to use a public key certification-based service between the mobile customer and the bank which will guarantee authentication and non-repudiation but given the restricted CPU power of mobile devices this is not feasible. The Diffie–Hellman session key and AES or 3DES encryption is normally used. The communication between WAP/Mobile Payment Gateway servers and Bank' servers must however be based on digital signature, public key certificates, as they use landlines. The communication channels connecting them should preferably be VPN and TLS encryption should be used for all communication. As we pointed out in Section 8.7.1 to ensure end to end security payment gateways must have agreements with mobile network operators to allow them to operate their own WAP gateways integrated with their servers (See Figure 8.12).

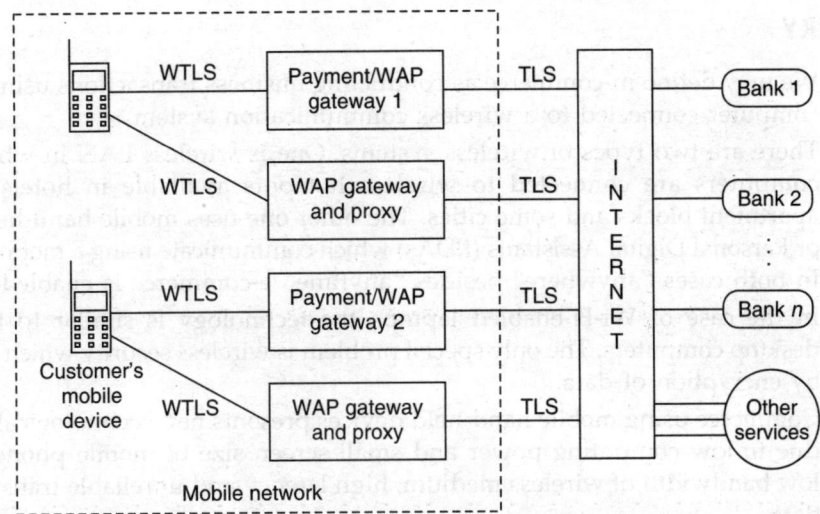

Figure 8.12 Mobile banking private payment/WAP gateways.

Observe that there may be several network operators with payment gateways with WAP gateway which could be connected to the servers of their subscriber banks by VPN/TLS. A transaction will originate from a customer using his or her password to gain entry to the gateway server followed by mpin to initiate banking transaction. All transactions between a mobile customer and the WAP gateway uses WTLS security. Beyond this point SSL/VPN is used. Needless to say that WAP servers, payment gateway servers and bank servers must all have firewalls and take all precautions to prevent intrusion by third parties.

RBI guidelines mandate end to end security from the mobile phone user to the bank. This is not possible in a literal sense as at the payment gateway there is protocol conversion from WAP to IP and this requires plain text availability for a few milliseconds as explained in Section 8.7.1. Thus, there must be trusted relationship between payment gateway and the bank covered by appropriate legal contract.

8.10 CONCLUSIONS

The number of mobile subscribers is increasing at about 8 million per month. With the advent of 3G services in India for which mobile operators will be paying large amounts for spectrum use they must find several value-added services besides voice service. Among these will be services such as audio, video on demand, mobile commerce, mobile banking, etc. Currently, the mobile hand-held device market is primarily small screen low cost devices (less than Rs. 5000). The increase in bandwidth when 3G is introduced would make response times for digital data exchange faster. However, the screen size limitation will remain. Larger screen, full function keyboard devices are now widely available which will be more convenient to use for m-commerce which requires good information browsing facilities. These mobile devices are currently quite expensive (around Rs. 25,000). It is expected that with the increase in customer base and reduction in cost of integrated circuits and display devices, these devices will become more affordable. New services are also starting with e-book availability on specially designed mobile devices (e.g., Amazon's Kindle). Thus, the future of m-commerce looks very promising.

SUMMARY

1. We may define m-commerce as conducting business transactions using a mobile computer connected to a wireless communication system.
2. There are two types of wireless systems. One is wireless LAN in which laptop computers are connected to wireless hotspots available in hotels, airports, apartment blocks and some cities. The other one uses mobile hand-held phones or Personal Digital Assistants (PDAs) which communicate using a mobile network. In both cases "anywhere" besides "anytime" e-commerce is enabled.
3. In the case of Wi-Fi-enabled laptops the technology is similar to the use of desktop computers. The only special problem is wireless security which is handled by encryption of data.
4. Commerce using mobile hand-held devices presents new technological problems due to low computing power and small screen size of mobile phones and the low bandwidth of wireless medium, high latency and unreliable transmission of data.
5. We discuss m-commerce by using a layered architecture. There are four layers in this architecture. The lowest is the wireless signal carrier known as "bearer", the next is the logical layer which consists of the Wireless Application Protocol Stack (WAP). This layer supports a services layer consisting of a microbrowser and a wireless markup language (WML) and a top layer of applications such as B2C e-commerce, m-payment, etc.
6. The unique advantage of m-commerce is anywhere e-commerce allowing users to transact business while they are on the move.
7. Applications may be classified as those using Short Messaging Service (SMS) of mobile phones, accessing services on mobile Internet, location dependent applications and those where mobile laptop computers are used.
8. Analogues to Internet Protocol stack which enabled the creation of the Internet, wireless hand-held device manufacturers and mobile network operators cooperated to agree on a common communication protocol for mobile devices called Wireless Application Protocol.
9. WAP-enabled mobile phones connected by mobile networks constitute a "mobile Internet". Mobile Internet is independent of the type of mobile network technology (GSM or CDMA) and brand of hand-held devices (Nokia or Motorola or Samsung, etc.)
10. WAP protocol consists of six layers: Application Layer (WAE), Session Layer (WSP), Transaction Layer (WTS), Security Layer (WTLS), Datagram Transport Layer (WDP) and the Physical Layer called bearers. Table 8.2 summarises the services provided by each layer.
11. In order to allow mobile hand-held devices to access the Internet, there is a need to convert WAP protocol stack to TCP/IP stack. This is done by a device called WAP gateway at the boundary of a mobile network.
12. Microbrowsers available with mobile phones should display information on a small screen in 3 or 4 lines. Thus, HTML cannot be used. A markup language known as Wireless Markup Language (WML) is used for mobile devices.

13. HTML should be converted to WML to display on mobile devices. This is done by a WAP proxy server. An alternative is for the web applications server to present information in WML format for mobile phones.
14. WML is organized as a set of linked cards with 3 or 4 lines of information per card. A card deck would be similar to a HTML page. Syntax of WML is similar to XML.
15. Security of wireless data is ensured by wireless transport security layer (WTLS). This is similar to TSL of the Internet. It uses RSA/3DES or Diffie–Hellman key exchange algorithm. It is slow as CPU power of mobile devices is limited.
16. During WAP to IP conversion plain text is used at WAP gateway for a few milliseconds. Thus, end to end security between mobile devices and web server is not available. This is alleviated by either WAP gateway being managed by the web service provider or shifting WAP gateway to web server using packet tunneling.
17. There are four payment methods. They are: debiting SIM card of a mobile phone, use of SMS for authorizing payment, using smart card with mobile devices and credit card payment initiated by a mobile phone which supports WAP.
18. Prepaid SIM cards in mobile phones are debited by network operators for providing services such as cricket scores once an SMS is sent. Similar SMS initiated requests to other vendors may be made which are billed by the vendor or debited from prepaid account with the vendor. These services are used for small payment below Rs. 1500.
19. A method similar to that used in the Internet-based credit card payment is used with WAP enabled mobile devices. SET protocol is not used as it is compute intensive and mobile devices have limited CPU power.
20. Mobile banking is becoming popular. There are two types of services. One is to obtain information such as account balance and last five transactions. The other is for funds transfer which changes the account status and requires better security.
21. Enquiries to banks relating to a customer's account are enabled by registering mobile number and sending an SMS. Funds transfer requires a separate mobile personal identification number coupled with a mobile personal account number and previously registered e-mail ID or phone number. Normally, a payment gateway which cooperates with the bank and the mobile network operator acts as an intermediary and ensures secure transactions.

EXERCISES

8.1 Define m-commerce.
8.2 What are the differences between e-commerce and m-commerce?
8.3 What are the unique technical problems which arise when implementing m-commerce?
8.4 Give a layered architecture for m-commerce. Give the functions of each layer. What are the advantages in describing m-commerce using a layering approach?
8.5 Enumerate the applications of m-commerce. What are the special advantages of m-commerce when compared to e-commerce? What are the disadvantages?

8.6 What services are usually popular based on SMS? What is the maximum length of messages which can be sent using SMS?

8.7 What are the advantages of Wi-Fi LAN-based systems in m-commerce as compared to mobile phone-based m-commerce? What are the disadvantages if any?

8.8 What is WAP? In what way does a WAP-enabled mobile phone differ from a voice-based mobile phone.

8.9 Enumerate m-commerce applications using WAP enabled phones.

8.10 For what applications can a salesperson use a WAP-enabled mobile phone?

8.11 Explain the term "location dependent m-commerce". Give two examples of location dependent m-commerce. Explain how a mobile device is used in the examples given by you.

8.12 What cellular phone technologies are widely used in India? Are they digital or analog? Are they compatible?

8.13 What is the general architecture of a cellular telephone system?

8.14 What is MTSO? What is a base station? What are their functions?

8.15 How is communication established between two mobile phones in a GSM system?

8.16 What is cellular structure in GSM? Why are cells hexagonal?

8.17 Why is WAP required in a mobile phone system?

8.18 Give the layered architecture of Wireless Application Protocol. Why is a layered architecture used?

8.19 Explain the functions of each layer in WAP stack.

8.20 Who are the stakeholders in wireless data services? What are the services provided by each of them?

8.21 What is a WAP gateway? Why is it required? What are its functions?

8.22 Compare the layers of WAP stack and that of TCP/IP. Does WDP correspond to UDP or TCP?

8.23 What are the special problems of mobile communications as compared to the Internet on land lines? How is this taken into account in WAP design?

8.24 Is HTML suitable as a language for displaying information on mobile hand-held devices? If not, give the reasons why it is not.

8.25 What is WML? What is the general structure of a WML deck? How many lines are normally allowed in a WML card?

8.26 In order to render information from web pages on a mobile hand-held device what transformation is required? What problems do you see in doing this?

8.27 Write a WML program to display

```
This display gives share prices
```
Screen 1

```
ABC  Rs.252
XYZ  Rs.385
PQR  Rs.585
```
Screen 2

8.28 Write a WML program to input data given below:

```
------------ INPUT ------------
Employee No: [      ]
Name:        [      ]
Dept:        [      ]
```

8.29 What is XHTML? In what way is it different from HTML?

8.30 What are the difficulties in implementing WTLS in mobile hand-held devices? What is the usual encryption method used in WTLS?

8.31 What special security problems are encountered when a WAP-enabled mobile device is to communicate with web sites in the Internet? Is end to end security ensured? If not, what is the reason? How can this be alleviated?

8.32 What are the various possibilities of placing WAP gateway? What are the requirements of each?

8.33 If WAP gateway is placed at the web site of an organization how is WAP stack sent to the gateway? How does it improve security?

8.34 There are two ways of converting HTML to WML. One is to have a general purpose software running on a proxy server, the other is to have the HTML content to be appropriately rewritten as WML document by the content provided. Discuss the relative merits and demerits of these two approaches.

8.35 Enumerate the different methods of mobile payments. Describe how an SMS-based payment system works.

8.36 How is SIM card used for mobile payments? Give some examples.

8.37 How is credit card payment implemented with a mobile hand-held device? How is the security problem solved? Give a block diagram showing different parties who participate in credit card payment and their roles.

8.38 What is mobile banking? What information services are provided in mobile banking?

8.39 How is security ensured in mobile banking information services?

8.40 How are mobile banking transactions involving transfer of funds implemented? What special security methods are used in such transactions?

OBJECTIVE QUESTIONS

Each question has four possible answers. Pick the most appropriate answer.

8.1 By m-commerce, we mean
   (a) Mobile e-commerce
   (b) E-commerce using mobile phones
   (c) E-commerce using laptop computers
   (d) E-commerce using portable computers connected to wireless LANs or using mobile hand-held phones/devices

8.2 Unique advantage of m-commerce is
   (a) Anytime e-commerce
   (b) Anywhere e-commerce
   (c) Anytime and anywhere e-commerce
   (d) Inexpensive e-commerce

8.3 Special problems encountered in m-commerce using mobile phones are:
   (i) The screen size is small.
   (ii) The keyboard has limited keys.
   (iii) Mobile devices are analog devices.
   (iv) CPU power is limited.
      (a) (i), (ii), (iii)
      (b) (i), (ii), (iv)
      (c) (i), (iii), (iv)
      (d) (ii), (iii), (iv)

8.4 Disadvantages of m-commerce with mobile phones compared to e-commerce are:
   (i) Mobile network has low bandwidth and high variable latency.
   (ii) Mobile networks are less reliable and packet losses are relatively high.
   (iii) Security of mobile networks is poorer.
   (iv) Mobile phones are not widely available.
      (a) (i), (ii), (iii)
      (b) (i), (ii), (iv)
      (c) (ii), (iii), (iv)
      (d) (i), (iii), (iv)

8.5 In the layered architecture of m-commerce described in the book, there are:
   (a) Six layers
   (b) Five layers
   (c) Four layers
   (d) Three layers

8.6 Among all mobile services for sending textual data, the most popular is
   (a) GPRS
   (b) SMS
   (c) IPV4
   (d) GPS

8.7 For using laptops in m-commerce, the easiest method is to use
   (a) Wi-Fi hotspots
   (b) GPRS cards
   (c) GSM systems
   (d) CDMA

8.8 For location based applications in m-commerce, it is necessary for the mobile device to have
   (a) GPRS enabling
   (b) GPS facility
   (c) SMS
   (d) WAP enabling

8.9 SMS based mobile services include
   (i) Confirmation of airline ticket booking
   (ii) Getting cricket score
   (iii) Accessing of e-mail
   (iv) Alerts of delayed flights
      (a) (i), (ii), (iii)
      (b) (ii), (iii), (iv)
      (c) (i), (ii), (iv)
      (d) (i), (iii), (iv)

8.10 Wireless Application Protocol is used
   (a) To create a mobile Internet
   (b) In B2C e-commerce
   (c) For location based services
   (d) For mobile payment

8.11 Wireless Application Protocol is the analog of
   (a) Simple Mail Transfer Protocol
   (b) Internet Protocol
   (c) MIME Protocol
   (d) UDP

8.12 The two cellular technologies used in India are
   (a) AMP and GPRS
   (b) GSM and CDMA
   (c) BSNL and Airtel
   (d) Reliance and Vodafone

8.13 The shape of the cells used in mobile communication system is a
   (a) Circle
   (b) Triangle
   (c) Rectangle
   (d) Hexagon

8.14 The primary purpose of a base station (BS) is to
   (a) Connect mobile phones to the PSTN
   (b) Send and receive signals from mobile devices within its range
   (c) Manage mobility
   (d) Connect mobile phones in different cells

8.15 The size of a cell depends on
   (a) Density of traffic in the cell
   (b) Cost of base station
   (c) Terrain of the cell
   (d) None of the above as all cells are of the same size

8.16 The major functions of Mobile Telecommunication Switching Office (MTSO) are:
   (i) Managing handoff of mobile devices from one base station to another
   (ii) Establishing connection between a mobile phone and a phone connected to PSTN
   (iii) Monitoring calls in progress and charge for use
   (iv) Establishing connection between mobile phones in different cells
      (a) (i), (ii)
      (b) (i), (ii), (iii)
      (c) (i), (iii), (iv)
      (d) (i), (ii), (iii), (iv)

8.17 In a mobile network, the number of MTSOs is
   (a) Only one
   (b) About ten
   (c) Several depending on coverage
   (d) Irrelevant

8.18 General Packet Radio Service (GPRS) is
   (a) Radio stations which can be heard on mobile phones
   (b) A modem to convert audio signal on mobile phone to digital packets
   (c) Packet-oriented data service for mobile devices in GSM systems
   (d) Packetized radio service on a mobile network

8.19 Short Messaging Service (SMS)
   (a) Guarantees delivery of short messages
   (b) Is a best effort message delivery system
   (c) Is available only in GSM phones
   (d) Allows messages only in home network

8.20 The maximum length of messages in SMS is
   (a) Unlimited
   (b) 160 characters
   (c) 80 bytes
   (d) 320 bytes

8.21 WAP protocol stack has
   (a) Four layers
   (b) Five layers
   (c) Six layers
   d) Seven layers

8.22 Wireless Transaction Protocol (WTS) in WAP provides
   (i) Reliable end-to-end message exchange
   (ii) Explicitly pairing of client request and server response
   (iii) Optimizes bandwidth use
   (iv) Allows efficient operation of microbrowser
      (a) (i), (ii)
      (b) (i), (ii), (iii)
      (c) (i), (ii), (iii), (iv)
      (d) (i), (ii), (iv)

8.23 The security layer in WAP (WTLS)
   (i) Uses public key encryption
   (ii) Defends against attacks such as replay attacks
   (iii) Is bearer network dependent
   (iv) Is similar to SSL of Internet
      (a) (i), (ii)
      (b) (i), (ii), (iii)
      (c) (i), (ii), (iii), (iv)
      (d) (i), (ii), (iv)

8.24 WAP gateway is required to
   (a) Convert session/transaction (WSP/WTP) of mobile network to session/transaction (HTTP) of Internet and vice versa
   (b) Convert the entire WAP protocol stack to Internet Protocol stack
   (c) Present web pages in a format appropriate for mobile devices
   (d) Connect a wireless system to a wired system

8.25 A WAP proxy is required to
   (a) Convert wireless security (WTLS) to wired security (SSL)
   (b) Convert Internet data to wireless data
   (c) Convert WML pages to HTML pages
   (d) Convert HTML to appropriate WML for mobile devices

8.26 Wireless Markup Language (WML) is used to display
   (a) Messages in mobile phones
   (b) WML cards on the mobile screen
   (c) Web pages converted to a format suitable for display on mobile device screen
   (d) Information from web servers

8.27 WML
   (i) Is similar to HTML designed for a small screen of mobile devices
   (ii) Consists of a deck of linked cards each card displaying 3 or 4 lines of information
   (iii) Is an application of XML 1.0
   (iv) Uses user-defined tags as in XML
      (a) (i), (ii)
      (b) (i), (ii), (iii)
      (c) (i), (iii), (iv)
      (d) (i), (ii), (iii), (iv)

8.28 WML has tags to
  (i) Link to an arbitrary card in a deck
  (ii) Display images of any format
  (iii) Accept information from mobile device user
  (iv) Display small tables
    (a) (i), (ii), (iii)   (b) (i), (ii), (iii), (iv)
    (c) (i), (ii), (iv)   (d) (i), (iii), (iv)

8.29 XHTML is a new language which
  (i) combines features of XML and HTML
  (ii) is used instead of WML in mobile devices with larger screen
  (iii) is case sensitive
  (iv) does not require DOCTYPE specification
    (a) (i), (ii)   (b) (i), (ii), (iii), (iv)
    (c) (i), (ii), (iii)   (d) (i), (iii), (iv)

8.30 Wireless Transport Level Security (WTLS)
  (i) Is applicable to mobile network only
  (ii) Is similar to SSL/TLS of Internet
  (iii) Uses RSA/3DES encryption but is very slow due to limited CPU power
  (iv) Is optional in WAP
    (a) (i), (ii)   (b) (i), (ii), (iii)
    (c) (i), (ii), (iv)   (d) (i), (ii), (iii), (iv)

8.31 When WTLS encrypted messages are sent to Internet server
  (a) End to end security is not assured
  (b) End to end security is guaranteed
  (c) Messages are converted to plain text partially
  (d) WAP gateway converts messages to plain text for Internet transmission

8.32 When WTLS is transformed to SSL/TLS for transmission on the Internet
  (a) WAP gateway just stores and forwards the secure packets
  (b) WAP gateway has to convert WTLS to plain text and then encrypt it in TLS to send it on the Internet
  (c) WAP gateway converts WTLS to TLS without accessing plain text
  (d) A special WAP proxy server is used

8.33 To ensure end to end security of information sent from a mobile client to a server on the Internet the methods used are:
  (i) Have secure WAP gateway
  (ii) Provide additional WAP gateway(s) which are under the control of web service provider(s)
  (iii) Send WAP packet as is on the Internet and perform the WTLS to TLS conversion at the web server end
  (iv) End to end security is provided when WAP gateway with firewall is used
    (a) (i), (ii)   (b) (i), (ii), (iii)
    (c) (ii), (iii)   (d) (ii), (iii), (iv)

8.34 VPN connectivity of a mobile client to a web server is not practical because
  (a) VPN connectivity is defined only for the landline Internet
  (b) VPN protocols are not applicable to mobile networks

(c) There is too much delay in mobile net
(d) CPU power of mobile clients is not enough to implement VPN connectivity

8.35 Types of mobile payments available for services are
(i) Debiting from prepaid SIM card
(ii) Payments initiated by SMS
(iii) Using smart cards accepted by some mobile devices
(iv) Using credit cards from WAP enabled mobile devices
(a) (i), (ii) (b) (i), (ii), (iii)
(c) (i), (ii), (iii), (iv) (d) (i), (iii), (iv)

8.36 SIM card debit-based payments are normally for
(a) Services provided by the mobile network operator
(b) Payment of utility bills
(c) Debiting/crediting bank accounts
(d) Post-paid services

8.37 By micropayments we normally mean payments up to
(a) Rs. 50 (b) Rs. 100
(c) Rs. 500 (d) Rs. 1500

8.38 Micropayment are normally made using
(i) Credit card
(ii) Smart card
(iii) SMS
(iv) SIM card
(a) (i), (ii) (b) (iii), (iv)
(c) (i), (iii) (d) (ii), (iv)

8.39 To allow fast and convenient viewing of web content by mobile users, it is necessary to provide
(a) A WAP proxy
(b) A high-end mobile device with larger screen
(c) At the server convert web content to WML format for viewing by any mobile device
(d) HTML to WML conversion software by the mobile device

8.40 Credit card payment using mobile is
(a) Insecure
(b) Uses SET protocol as in Internet-based system
(c) Uses WTLS/TLS security and use the Diffie–Hellman key exchange algorithm
(d) Uses special WAP gateway which is security enhanced

8.41 By mobile banking we mean
(a) Sending an e-mail message to a bank
(b) Banking transactions initiated by a customer using a mobile hand-held device and a mobile network
(c) Anytime, anywhere banking
(d) Sending and receiving SMS to banks using a mobile phone

8.42 In mobile banking, several independent identifications are needed. Normally, they are
  (a) Mpin, registered mobile phone number and account number
  (b) Registered mobile phone number and account number
  (c) Password, mpin and account number
  (d) SIM card number, mpin and account number

8.43 Mobile banking transaction which affects the customer's account balance in a bank requires the following entities:
  (i) Reserve Bank of India
  (ii) Clearing entity for inter bank mobile transactions
  (iii) Mobile payment gateway operators
  (iv) Mobile network operators
    (a) (i), (ii), (iii)
    (b) (i), (ii), (iv)
    (c) (i), (iii), (iv)
    (d) (ii), (iii), (iv)

8.44 Mobile payment gateways
  (i) Provide technical services to a bank
  (ii) Ensure end to end security of transactions
  (iii) Guarantee settlement of funds
  (iv) Comply with regulations of Reserve Bank of India
    (a) (i), (ii)
    (b) (i), (ii), (iii)
    (c) (i), (ii), (iii), (iv)
    (d) (i), (ii), (iv)

8.45 In mobile transactions involving money transfer security is ensured by
  (a) Using digital signature of customer
  (b) WTLS to payment gateway and TLS with digital signature of gateway operator on the VPN to the bank
  (c) Diffie–Hellman encryption between customer and bank
  (d) RSA/3DES encryption between customer and payment gateway

8.46 General Packet Radio Service (GPRS) provides
  (a) FM Radio Connectivity to mobile phones
  (b) Digital data packet service in GSM mobile networks
  (c) Analog communication using wireless
  (d) Digital packets on CDMA network

8.47 General Packet Radio Service is used to provide Internet connectivity to
  (a) Mobile laptops
  (b) WAP-enabled mobile phones
  (c) High-end mobile phones using a GSM network
  (d) Mobile laptops using a CDMA network

8.48 CMDA 1x EVDO
  (i) Can be used by both laptops and high-end mobile phones
  (ii) Is a packet communication system provided in CDMA networks
  (iii) Can provide Internet connectivity to laptops using any wireless mobile network
  (iv) Provides Internet connectivity to laptops using a CDMA network
    (a) (i), (ii)
    (b) (i), (ii), (iii)
    (c) (i), (ii), (iii), (iv)
    (d) (i), (ii), (iv)

CHAPTER 9

E-Commerce of Multimedia

LEARNING GOALS

In this chapter, we will learn:
1. How the emergence of new methods of storing and distribution of digitized multimedia affect their marketing.
2. E-commerce of print publications in electronic form.
3. E-commerce of audio and video e-publications.
4. Intellectual property issues arising out of e-distribution of multimedia.

9.1 INTRODUCTION

Publication of multimedia which include books, magazines, music and movies is a huge industry with a total market of billions of rupees. Traditionally, the manufacture and sales of these goods used bulky material such as paper, magnetic tapes, magnetic disks and optical disks (CDs and DVDs). The entire field has been revolutionized by the advent of computers which facilitate production of multimedia in digital electronic format.

Distribution of digital electronic versions of multimedia is now possible directly from the producer to the consumer using the Internet and high speed communication networks. Interestingly, the first successful e-commerce application was marketing of printed books by Amazon.com. Amazon sells books by creating a huge virtual bookstore on their web site.

Amazon later expanded to audio, video and a variety of other products. Even though the sale and collection of money for the sale of books uses the Internet, the actual book is delivered by mail/courier. This model is currently undergoing a revolutionary change. The book itself which is available in digital form is being sent by Amazon electronically using a mobile/wireless network. For a consumer to read the book Amazon is selling a device called *Amazon Kindle* which is portable and uses e-paper making it easy for users to read and download books using a mobile network. Amazon is not only selling books but also newspapers, magazines, etc., at a cost which is much lower than their print cost. This will change the print publication industry significantly. Several new issues crop up, particularly Intellectual Property Rights (IPR) which should be solved in innovative ways.

So far the major impediment in e-publication and e-distribution of print material was the inconvenience of reading them using current display devices. With devices such as Kindle which is expected to emerge in coming years, this impediment will be alleviated and we expect an explosion of e-publishing and consequent emergence of new e-commerce business models in book trade.

As we pointed out earlier, publishing is not restricted to print material. Another big publishing industry is music. In this industry revolution came earlier with the emergence of digitizing and audio compression which resulted in MP3 compressed audio resulting in easy download using the Internet. The music industry felt threatened as a lot of music was being exchanged/downloaded without payment to music publishers. IPR law was amended in USA to prevent illegal distribution of copyrighted music. Here again a revolution is taking place with the emergence of ipod, a mobile device which can receive music and play it. Apple which makes this device sells music tracks which can be received on this mobile device and pays royalty to the publishers. Many record labels are now allowing music downloads on payment using the Internet.

Video on demand is already available in several countries and has now been introduced in India in a limited way by some Direct to Home Satellite TV companies. The satellite TV companies broadcast recent movies (without any advertisements interrupting them) to homes at fixed times on payment. Video on demand would allow customers to choose from a large library of movies stored online in a disk storage server by the provider and view them on payment using their home TVs, uninterrupted during times of their choice. The major impediment for such a service was the telecommunication cost and storage cost. Disk storage size is doubling every year with no increase in cost. Bandwidth is doubling every 9 months with no increase in cost. With set-top boxes on most TVs in urban areas Video on Demand service is bound to emerge as a viable e-commerce product soon.

E-commerce in multimedia in digital form has the unique characteristics that the items sold can be sent by using a communication system. The items may be ordered and paid for by using the Internet as in normal e-commerce. The product instead of being physically shipped is sent electronically using a separate communication system appropriate for the product. The Internet may also be used, in theory, to send the product but currently it is more expedient to use a separate communication system appropriate for, books, music and video. The nature of products which are termed "Intellectual Property" and the nature of distribution give rise to new technical and legal problems. Thus, the primary objective of this chapter is to describe the new technologies used in distributing multimedia, intellectual property protection and new business and payment models for multimedia which can be directly marketed by the publishers to consumers.

9.2 E-PUBLISHING OF MULTIMEDIA

We may define e-publishing as producing of books, periodicals, newspapers, music and video using computers rather than traditional media such as paper, magnetic media and photographic films and distributing the material using a telecommunication network rather than traditional distribution channels such as postal system, distributors and retail stores. In e-publishing all published materials will be stored in vast on-line storage devices attached to a server (as bits). The material will be distributed using telecommunication networks, either wired or wireless. There are two methods of using a telecommunication network. One of them is to use private networks to transmit appropriate published material to distribution centres. These centres can in turn transmit them using private networks (such as TV cable network or satellite system) to consumers. The other is to use the Internet and directly market the material to consumers.

There are three major categories into which we will divide publications. They are:

1. Those using paper and printing. We have in this category books, scholarly journals (i.e., science, technical and medical), popular magazines (e.g., India Today, The Week, etc.) and newspapers. This is the largest group of publications with perhaps the highest revenue.
2. Music of all types traditionally published by recording industry and distributed nowadays using audio CDs.
3. Video of all types which include movies, television programmes, live concert recordings and sports recordings published by movie industry and distributed using DVDs.

Printed material is static, i.e., the material need not be consumed (read) with no breaks. In other words, you can read a few pages of a book today and continue to read it a few days later. Audio and video, on the other hand, are time varying. In other words, time is an important variable in these cases. We hear music and see movies continuously in real time (preferably) with no breaks.

A general model of e-commerce of e-multimedia will consist of the following:

1. A repository of the material published and to be marketed. This repository will normally be stored in hard disks in several logical volumes. Disk stores would normally be organized with cache storage (semiconductor memory) to allow continuous downloading of material by consumers.
2. A distribution network which will be a telecommunication system appropriate for the type of material being distributed and whether it is sold or it is sent for viewing in real time as in a movie or a live concert to a TV set or home theatre.
3. The device for the consumer to use the product conveniently. For audio and video traditional systems such as hi-fi systems and TVs are appropriate. For books, however, reading on a video monitor of a computer is not convenient. The emergence of new e-paper like devices (using what is known as e-ink technology) has made e-distribution of print material practically feasible. We will now see how print material, audio and video are digitized and stored.

9.3 DIGITIZING AND STORING OF BOOKS

Print publications can be classified into two major categories in terms of digitizing technologies. Those which were published after the advent of computers and word processors and those before. All books published after 1995 and current publications fall in the first category. They are already in digital form and storable using standard formats such as portable document format (.pdf) or .doc format. If we assume a book to be 500 pages long with 45 lines per page and 80 characters per line, the total storage needed for storing the characters in the book = $500 \times 80 \times 45$ = 1.8 MB. If the book is a technical book and we assume 10% of the print area to be images, there will be 50 pages of images. If a page print area is ($6" \times 8"$) and we assume (1000×10000) pixels per inch resolution for good quality image, the number of bits needed to store 50 pages of images will be ($6 \times 8 \times 1000 \times 1000 \times 50$) bits = 300 MB. If we compress the images using jpeg format, the storage will be reduced nearly fifteenfold giving storage needed as 20 MB. Thus, (text + images) for a 500-page book will be ($0.9 \times 1.8 + 20$) = 21.62 MB. We thus see that most of the storage will be used up for images. A book of fiction, on the other hand, will have hardly any picture and will need only 1.8 MB for a book of 500 pages. If a publisher of technical books has 10000 books in print, the total storage needed to store all books in his catalogue will be 21.6×10000 MB = 216 GB which is not too large by today's standard.

The situation is quite different when older books are considered. Only print version in bound form exists for most of them. Particularly if we want to digitize books in a library or digitize old newspapers and store them for archival purposes, then we are forced to scan them using what are known as scanners. Fast book scanning is possible with new scanning technology in which a book is placed on a V-shaped scanner and imaging system is used to convert the information to picture elements or pixels. If the print material is black and white one bit is needed per pixel. After scanning, the page is turned automatically by a robotic arm to scan the next pages. Such expensive scanners are used by companies such as Google and Microsoft which are in the process of digitizing the contents of several libraries. Their objective is to use such a repository in their search engines to make retrieval by such engines more relevant. Several issues of Intellectual Property Rights arise in these projects and we will discuss these later in this chapter. We will now calculate the storage needed to store scanned books in a database.

Assuming a book with text area of ($6" \times 8"$) per page assuming the scanner has a resolution of (1000×1000) bits per square inch, the number of bits per page will be ($6 \times 8 \times 1000 \times 1000$) bits = 6 MB. A 500-page book if scanned and stored in bitmap format will be 3000 MB or 3 GB. Even with compression it will be 300 MB per book. It is too large. If on the other hand, the page is stored in ASCII format, the space will be reduced considerably. Another major advantage of storing text in ASCII form is that it can be indexed using the keywords in the text which enables search and retrieval. Assuming that a $6" \times 8"$ text has 80 characters per line and 45 lines per page, the number of ASCII characters per page is 3600 which is 3600 bytes. This is 600 times lower than a bitmapped page. Thus, it is essential to convert bitmap (or pixels) of images to ASCII characters. This is done by a software called Optical Character Recognition (OCR) software. The conversion is possible only if standard fonts are used in the text. Even then it is not 100% accurate. Accuracy of 99.5% is obtained with recent OCR conversion software. Thus, manual proofreading for correction is needed. The total storage needed for a book of 500 pages is 1.8 MB. We have assumed a book with no graphics. If 10% of the book has images which are to be stored in bitmapped from, the storage per book will be 301.6 MB as images will need 300 MB storage in bitmapped from. If images are compressed using compression algorithm and stored in .jpeg form, the image storage will be reduced fifteen fold and may be stored in 20 MB. Thus, the total storage for a book with images will be 21.62 MB. If an e-publisher has 5000 older books in print the storage needed will be 108 GB, not too large by current standards. However, the manual effort required for proofing the text, placement of images and organizing the book will be considerable. The total storage for old and new books would be 108 + 216 = 324 GB. For a bookseller such as Amazon who will be selling books of several publishers, the situation will be quite different. If publishers give Amazon access to e-books to sell in retail, it still has to organize them in their own database for easy and fast retrieval. Currently, they have a collection of around 15000 e-books. Assuming e-books are plain text with no images and an average size to be 500 pages each book will need 1.8 MB. A much larger collection of 150,000 books will need only 270 GB. To facilitate fast retrieval and distribution to customers the collection will be stored in several logical volumes and may be even duplicated and stored in several physical disks. This topic is usually discussed in books on textual database organization and is not in our scope.

9.4 DIGITIZING AND STORING AUDIO

Audio signals acquired by microphones is in analog form (i.e., continuously varying waveforms). They are nowadays digitized and stored. Digitization improves fidelity. Digital form can be compressed using a system called MP3. MP3 is the short form for Moving Picture Experts Group Version 2–layer 3 audio compression standard. This compression method has been standardised by an international group of experts from industry, government and academia. The compression algorithm uses the fact that when both louder and softer sounds are mixed people tend to hear the louder sounds. During digitization of the audio signal, the number of samples may be reduced during periods when its frequency is low. Intermediate values of an audio signal may be interpolated from the surrounding values. These methods of reducing the number of samples reduces the storage requirement of digitized audio signals by a factor of 10 to 14. A 60-minute music CD in compressed MP3 form will need around 50 MB. If a music publisher has 100000 CDs, the company can store them using MP3 format in 5000 GB. This is not too large by current standards to be stored in disk storage online. The music will be indexed and spread over several disks to expedite retrieval. A current business model is to market individual tracks of a CD and not the entire CD. Normally each track may be around 5 minutes. Each track has to be indexed. The index itself will need fair amount of space. The music will thus be distributed in several disks in a properly indexed form to allow individual tracks to be downloaded fast. The disk system will normally include a cache (a fast RAM) to expedite retrieval.

A e-music store may not store all its stock of 100,000 CDs online. It would normally store 10,000 popular CDs for online distribution which is around 500 GB. It should organize the music properly indexed for easy retrieval.

9.5 DIGITIZING AND STORING VIDEO

Video consists of succession of pictures. There are around 30 pictures per second so that with persistence of our vision, we see a moving picture. Each picture consists of a collection of picture elements (pixels) displayed in two dimensions. A standard video frame should have at least 640 × 480 pixels with 3 bytes per pixel for colour pictures. Thus, a 1 hour video with 30 pictures per second will require storage of 640 × 480 × 3 × 30 × 60 × 60 = 99.5 GB. Normally, Indian movies require 3 hours. Thus, one 3-hour movie will need approximately 300 GB. 1000 movies will need 300 TB, too large even by today's standards. We thus require the movie to be compressed, before they are stored. Compression is also necessary to transmit a movie using a communication medium to a recipient as the speed of communication medium is limited. The recipient of a compressed movie will expand it with a processor in his TV for viewing. We will now see how video is compressed.

There are two major principles used to compress movies. Firstly, each picture can be compressed using a compression algorithm used in jpeg compression. Secondly, from one video frame to the next only a few pixels change. Thus, if we take a group of pictures we need to store only changes between pictures and not all the pictures. These principles are used in algorithms developed by standardization group called Motion Picture Expert Group. The early most commonly used standard was MPEG2. A new standard called MPEG4 standardized as H.264 by an International body is now gaining acceptance as it gives a higher compression without loss of quality. The actual possible compression is variable depending on the nature of the video. The compression is by a factor of between 100 and 150 in MPEG2 and between 200 and 300 in H.264. Thus, 3-hour movie MPEG4 (H.264) compressed will require (300/300) GB = 1 GB. If a movie distributor has a library of 5000 videos, the storage needed to store them will be 5000 GB or 5 terrabytes (5 TB) which is not too large.

9.6 DISTRIBUTION OF E-BOOKS

In Section 9.3, we saw how books are digitized and stored in disk stores. In order to market these books, the following actions are required to be carried out by the bookseller/publisher:

- An e-book catalogue should be maintained by the e-bookseller.
- A sample chapter and reviews of the book (if any) should be available to a customer if requested by him or her.
- The e-books themselves should be digitized, properly indexed and stored in a database. The database should normally be a distributed database stored on several disks with requisite caches to enable fast download even when several customers simultaneously request the same book.
- Every customer should own an e-book reading device such as Kindle, Sony e-book reader, etc., which uses e-paper and is portable, light and battery operated. It should also have local storage to store several books.
- The e-book reading device should preferably be connected to a broadband wireless network using which the e-book seller could transmit the selected books. Normally, users of e-books prefer portable light weight battery operated mobile devices. This implies wireless transmission. As transmission is primarily bit strings GPRS or CDMA 1x DVO would be appropriate. Amazon's e-book reader Kindle is a mobile device which uses CDMA 1x DVO.

The procedure of obtaining e-books from a bookseller/publisher will be:

Step 1: A customer logs on to the website of the bookstore and examines the catalogue of e-books.

Step 2: He or she may select a book and request to see a sample chapter and book reviews (if any).

Step 3: If he or she decides to buy, he or she clicks "buy" in the browser. The bookseller requests the customer's credit card details and payment amount.

Step 4: Once the payment is approved, the e-bookseller will retrieve the book from the e-book database using the ID of e-book and transmit it to the customer's e-book reader using a mobile communication network.

Other interesting aspects of an e-book reader are:

1. The battery power is consumed by e-book only when pages are turned. While reading a page no battery power is used. This allows the battery to be recharged once in two days with normal use.
2. As the customer's e-book reader is the only device in which he or she can get the book downloaded and the data is digital, data packets may be sent using Internet Protocol.
3. IPR problems can be solved by encrypting the book using an ID specific to an e-book such as its serial number and decrypting with a built-in hardware decrypting device in the e-book reader.

9.7 DISTRIBUTION OF AUDIO

In Section 9.4, we saw how audio is digitized and stored in a repository by e-publishers/distributors of music. A business model popularized by Apple is to sell music tracks rather than entire albums. This gives freedom to customers (who are mostly youngsters) to spend smaller amounts and pick their favourite track. Entire albums may be sold (may be of live concerts) and appropriately priced. There are five components in this system. They are:

1. A customer with Internet access to a e-music shop's web site.
2. Music shop's web site which has a catalogue of music tracks/albums with prices. The shop may also send partial music tracks for listening by a customer to make a choice.
3. A database of audio in MP3 compressed form.
4. A temporary store connected to the database where retrieved tracks are put for being sent to a customer.
5. A customer's mobile audio device to which music tracks are sent for storage. Alternatively MP3 tracks may be sent using the Internet to the customer's computer.

The steps followed are similar to the procedure used for e-books:

Step 1: A customer logs on to the web site of a music store and examines the catalogue of tracks. He or she may ask for samples of tracks to hear on his or her computer before ordering. Sample MP3 files are sent by audio shop for the customer to listen. The customer selects tracks he or she is interested in and orders them.

Step 2: The e-audio shop sends a bill.

Step 3: The customer normally would pay the billed amount using a credit card. Else a payment method for information goods explained in Chapter 6 may be used.

Step 4: Once the payment is received, the audio shop sends a command to its music database server to retrieve the track(s) using track ID.

Step 5: The server retrieves the track(s) and transmits them to a temporary storage for onward transmission to the customer and informs the audio shop's computer.

Step 6: The audio shop's computer sends a command to the store to transmit the music track either wirelessly (mobile system is common) to the customer's audio device such as ipod or to customer's computer.

We have assumed that MP3 compressed audio is used for storage and transmission.

9.8 VIDEO ON DEMAND

Unlike e-books and audio which are purchased, movies/videos are normally rented and viewed. The system which is currently available (not yet in India) is called *Video on Demand*. In India some satellite TV providers broadcast recent movies at specified times which can be viewed by home viewers on payment. Normally, the payment is by authorizing the provider by SMS to debit prepaid accounts of customers. Video on demand allows viewing of any of the movies/videos in the library of a provider at anytime chosen by a customer on his or her home TV. The charges would depend on various factors determined by the service provider.

A video on demand system consists of the following parts:

An ordering system

A home computer of a customer using which he or she can log on to the provider's web site. The web site will have a catalogue of available videos with prices. The pricing would depend on whether the movie is current, whether it is popular, requested time for viewing, etc. A customer may order a video to be viewed on his home TV. Thus, the ordering is just like ordering for any goods/services. Payment would normally be by debiting a prepaid account of a customer maintained with the vendor.

Video storage system

The video on demand provider will have a library of video material stored in an online disk system. We saw in Section 9.5 that a library of 5000 videos each of 3 hours duration will require 5 TB. Whatever video is to be shown on demand should be online on a hard disk. The provider may organize its storage into a hierarchy of DVDs (which can be mounted on a reading system with a programmable robotic arm), an online fast hard disk and a large semiconductor cache memory. The primary objective of hierarchical organization is to minimize the cost of the video storage system without compromising on the real-time continuous viewing quality of videos provided to customers. The design of such a hierarchy is technically challenging and outside our scope of discussion. We will assume a repository of movies in compressed form is available from which videos can be delivered jitter free, in real-time, on demand to several customers.

Video server

The video storage system has an attached computer called a video server. The video server has the following functions:

Admission authentication: When a customer requests a service, the server should check user's password and credit availability before allowing access.

Admission control: When a customer requests a video, the CPU has to check if it can be scheduled immediately. This decision would be based on how many customers are already watching the video and buffer space availability. If a customer cannot be admitted immediately one may intimate him or her the earliest time when his or her request can be honoured.

VCR functions: They allow customers to mimic VCR functions such as fast forward, fast reverse, jump to specified point, etc.

Guaranteed stream transmission: Once a video is started for a customer, there should be no interruptions. The video should be sent at a steady rate. This guarantee can be given only if the network bandwidth is reserved and playback from video storage is sustained.

Encryption of multimedia stream: The multimedia data should be encrypted to prevent unauthorized access to the service. Encryption uses unique ID of STB (Set-top Box).

Billing and accounting: The server should have a program for computing amount payable by customers, debit prepaid amount and keep track of payments. It should also display current balance in a user's account when requested.

Quality of service: Facilities should be available to monitor quality of services provided to customers and log customer feedback.

Information service: It must have a list of videos available, their cost, synopsis of the story, type of video (e.g., language, comedy, musical, drama, etc.) which should be displayed on a customer's TV on demand.

Set-top box at customer's premises

At the customer's premises there will be a TV with an extra device called a Set-top Box (STB). The STB converts MPEG4 compressed digital video signal transmitted via a distribution network to uncompressed video and converts the digital signal to a format appropriate for the TV at the customer's residence. As the video signal is encrypted STB will decrypt it. It will also have a facility to accept commands from a hand-held remote control to pause, rewind, fast forward, etc. It must have enough buffer memory to store portions of a video to pause, rewind and replay for a short duration. The STB may also provide information on recent additions to the video library, current balance in customer's account, a gist of a movie's story, type of movie (comedy, drama, etc.).

An alternative method of ordering video would be by using the set-top box which is in fact a small special purpose computer. In this case, the remote control will be designed with user-friendly menu buttons for ordering besides those needed to control the video. TV screen will be used in a way similar to VDU of a computer. The remote control can be used to view a catalogue of videos with their prices besides information on duration of video, type of video, story, etc. A customer will be able to choose a video and request immediate delivery or request for delivery at a time chosen by him or her. Charges will be debited from his or her prepaid account and the account balance displayed on request. This is a good method for persons who either do not have a computer or are averse to using one and find it easier to use a TV remote with a user-friendly menu.

Distribution network

A distribution network will consist of cables connecting the vendor's site with all its client sites. The cable distribution network must have a bandwidth of at least 2 MBps for each client to distribute compressed video without distortion. The distribution network changes with change in technology. Technologies such as video for mobile devices and video over IP are emerging. The VOD provider will change its distribution system to suit its business model. There will be a communication path from STB to the VOD provider's system to accept user commands. This path will have much lower bandwidth as only commands are to be sent.

In Figure 9.1, we show the components of a video on demand system.

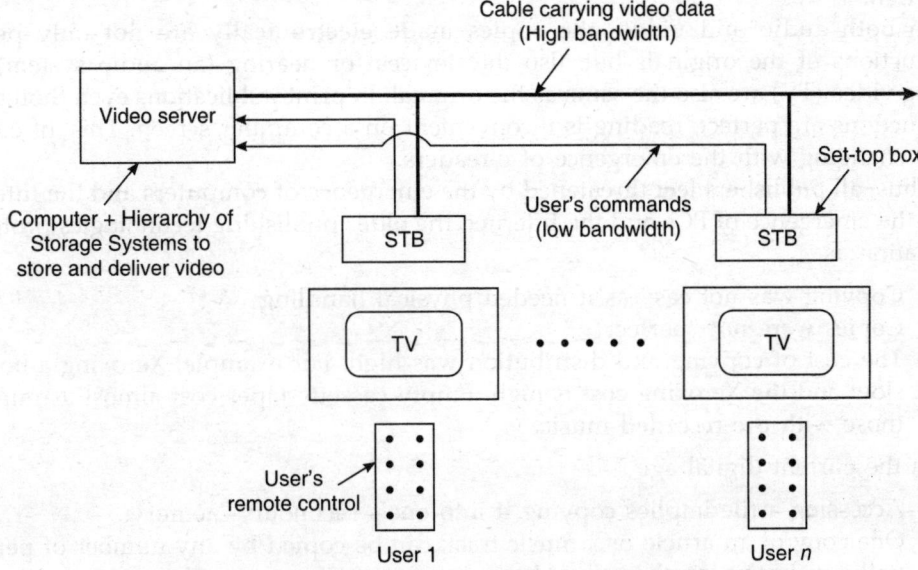

Figure 9.1 Components of a video on demand system.

9.9 INTELLECTUAL PROPERTY ISSUES

Traditional publishers of books, music and movies have reservations about e-publishing. The problem first arose in music business. A company called Napster invented a so called peer to peer music sharing program. This program allowed an individual "A" with music tracks stored in his or her PC to be accessed by another person "B" from his or her own PC using their Internet connection. It is called a peer to peer file sharing program. It assumes that any two persons connected to the Internet mutually agree to allow certain files to be copied between them. If a person "A" wants a particular music track, he or she will not know who has it and whether he or she would allow the music to be copied. In order to facilitate sharing files, napster.com maintained a directory giving the site addresses and the tracks available on these sites to copy. Person "A" can log on to napster.com and find out who has the music track he or she wants, log on remotely to his or her PC and transfer the file. The web site napster.com became so popular that the music industry felt threatened as a large number of recently published copyrighted music were being exchanged without any payment to the music publisher. A case was filed in the USA court which opined in favour of the music publishers and ordered napster.com to shut down. Napster.com argued that they were not *distributing* copyrighted music without paying any royalty to the publisher, but were only maintaining a directory and the individuals who allowed copying were the culprits. However, this argument was not accepted by the court and they opined that napster.com was an accessory to enable others to copy illegally. Napster.com was shut down but reappeared later as a music distribution site and paid royalty to publishers.

A similar problem arose in video with the emergence of a company named Bittorrent which developed a peer to peer video sharing program. Video as we have seen requires much higher bandwidth for communication compared to audio. Bittorrent's contribution was to expedite transfer of video files on the Internet between peers who have broadband connection.

In both audio and video, the copies made electronically are not only perfect reproductions of the originals but also the devices for hearing (an audio system) and viewing video (TV) are also the same as the original. In print publications even though the reproductions are perfect, reading is inconvenient on a computer screen. This, of course, is now changing with the emergence of e-readers.

Thus, all publishers feel threatened by the emergence of computers and the Internet. Before the emergence of PCs and the Internet, the older publishing technologies protected information as

- Copying was not easy as it needed physical handling
- Copies were not "perfect"
- The cost of copying and distribution was high. For example, Xeroxing a book is slow and the Xeroxing cost is high. Empty cassette tapes cost almost as much as those with pre-recorded music.

In the current digital age

- Accessing a file implies copying it into one's secondary memory.
- One copy of an article or a music track can be copied by any number of persons without destroying the original copy.
- Copies are perfect reproductions indistinguishable from the original.
- Cost of reproduction is very low.
- Digital files can be distributed instantly at negligible cost.

Thus, the natural barriers to copyright infringement of an earlier era are now removed. Further copies can be distributed to almost every home with Internet connection and the behaviour of individuals in the privacy of their home is difficult to monitor. Thus, Intellectual Property Rights (IPR) infringement is difficult to detect and track.

IPR law (also known as copyright law) historically tries to balance public good and private gain by

- Allowing public access to a limited number of copies in public libraries without paying royalty to publisher.
- Providing IPR to authors/publishers for a limited period (currently 60 years).

IPR provides the following rights to an author:
- Give a performance based on the work if it is music/drama.
- Exhibit the work if it is a work of art.
- Translate the material to any language.
- Make a movie based on the work (if it is, for example, a novel).

If anyone violates the above rights of the author without written permission, it is a criminal offence and can be punished with a jail term or fine or both.

IPR laws are country dependent. There are, however, International agreements on copyrights and their validity.

With the advent of electronic media, copyright protection has become a challenge. Some countries, particularly USA has tilted their balance overly to protect the intellectual property of music producers and movie producers who have a strong lobby in the US Congress. In USA, a law called Digital Millennium Copyright Act was passed in 1998 which made even attempts to develop algorithms to decrypt encrypted data an offence punishable with a jail term. They also amended an earlier law which allowed production of certain types of derivative works based on an existing work without paying royalty. For example, if a company selects from a telephone directory all doctors and publishes a directory of doctors, it was allowed. The new law prohibits this. The new law prohibits using parts of a number of copyrighted publications to create a new one. For example, music remixing without the permission of the original copyright owners is an offence.

In spite of the strong IPR laws, enforcing these in the new digital age is a challenge. Thus, IPR holders have developed two strategies to protect their intellectual property. They are:
- Technical solutions
- Business solutions

Technical solutions are:
- Use an encryption key which is sold/sent with the material. For example, a book sent to an e-book reader would be encrypted with encryption key which is specific to an e-book reader.
- Material may be read but not copied. Similarly, CD may have a read only coding and will not allow copying.
- Use digital watermarks. These are hidden symbols which are embedded in the material by watermarking algorithms. The watermark can be detected by a program designed by the creator of the material. Illegal copies will not have the mark and can thus be detected and used to punish the copier.

Technical solutions are not always immune to copying. Clever hackers find various means of breaking the system taking it as a challenge. Breaking protection mechanism is of course illegal.

Understanding that in the new digital age technical methods of copyright protection are not always effective, many publishers have innovated business solutions which are often more effective.
- License the use of the product rather than sell it. If a CD is sold it can be legally copied for own use or back-up. Used CD can be sold as a person owns it. In fact,

second-hand book sale is a thriving legal market where the publisher/author do not get any royalty. If a product is licensed, it cannot be sold in the second-hand market. In some cases, the licences have limited validity (particularly software) and at the end of the period it is programmed to self-destruct when read (including copies). Another model is to support the product only for the licensing period (particularly in software products).

- Make it easier and cheaper to buy e-copies than copy. For example, in many music CDs which have several tracks only certain tracks are popular. A buyer is forced to buy a whole CD. Apple came with the innovative solution of selling single tracks for $0.99 and deliver them electronically to the hand-held mobile device called ipod. This became very popular as the amount paid is low and delivery of the product is almost instant. Similarly, Amazon's e-books cost much lower than their print versions.
- Customize for individuals. For example, select short stories of several authors and make a customized anthology. Similar collections of music tracks may also be created on demand.
- Give away product free and make money on related products. For example, many musicians make some of their music available free on the Internet. Once they become popular they make money giving live concerts.

The following business strategies have been used by software product companies. In most countries software is protected by copyright.

- Give the copies at a low cost. Depend on volume sale and constant updates. For example, a compiler for Pascal called Turbo Pascal was sold for $49 depending on volume sale to make a profit. They also constantly improved the compiler encouraging further sales.
- Give away the software free. Make money on services. Linux is an open source free software package. Red Hat Linux makes money for support and services provided by it for Linux operating system software. This model is also used by some products. For example, Satellite Radio sets are very inexpensive. They may cost only Rs. 1500. However, the yearly subscription to receive satellite radio stations is Rs. 1800.
- Add value to legal subscribers by providing several extra features on the web. For example, several print magazines such as *The Economist* have excellent material on the web accessible to subscribers.

We conclude this chapter by observing that there are "soft goods" which are currently available in digital form such as books, music, video and software for computers. E-commerce in these goods is distinct as they can be distributed in digital electronic form almost instantly. (It is not only these, but also others such as airline e-tickets which are distributed electronically). Copying these goods is easy and so protecting them from thieves is a challenge. Even though laws are there to protect intellectual property rights, they are difficult to enforce in this digital age. Thus, innovative business solutions should also be found to make the risk of copying not worthwhile for most legitimate users of these goods.

SUMMARY

1. Multimedia products include books, music and video. Publishing multimedia is a huge industry.
2. With changes in technology multimedia products are now in digital form (bit strings) and they can be sold either by e-stores or by publishers directly to a customer and sent to the customer instantly (within a few seconds) using digital communication media (either cables, Internet or mobile communication).
3. Storing vast quantity of multimedia by an e-publisher has now become feasible as disk storage is doubling every year with no increase in cost. Distribution is also becoming feasible using communications media as communication bandwidth is doubling every 9 months with no increase in cost.
4. Books were not convenient to read on VDU. However, with the emergence of e-paper and mobile portable readers books can be distributed electronically.
5. Music in MP3 compressed form may be sold and distributed electronically.
6. Unlike books and music, movies (i.e., video) are normally rented and payment is for viewing. Video on demand systems allow customers to download video from online storage of the provider at any time and view it on their home TV sets on payment.
7. A serious problem with digital electronic multimedia publishing is the difficulty of protecting intellectual property rights of publishers and authors/musicians/actors. With the emergence of the Internet, it is very easy and inexpensive to distribute multimedia. Further, digital copies are indistinguishable from the originals.
8. Even though there are laws protecting intellectual property, they are difficult to enforce.
9. Businesses use technical means such as encryption and watermarks to protect their products from illegal copying.
10. Technical methods are mostly but not always effective in protecting intellectual property. Thus several e-publishers adopt business strategies which are more effective in discouraging copying multimedia illegally.

EXERCISES

9.1 Define multimedia.
9.2 What is the major difference between e-commerce of physical goods and e-commerce of e-published material?
9.3 What is an e-paper? Find out about e-paper from wikipedia.org. What are the advantages of e-paper displays compared to LCD/TFT displays?
9.4 Find out about Kindle from the Amazon.com web site. How does Kindle receive e-books? How many books can be stored locally in its storage? Compare the prices of print versions and their e-versions of some books from the Amazon.com web site.
9.5 At what rate has the size of disk storage increased at constant cost?

9.6 At what rate has the bandwidth of communication networks increased at constant cost?

9.7 In what way has the increase in size of disk and bandwidth of communication system influence the growth of e-commerce in e-publications?

9.8 Define e-publishing.

9.9 What infrastructure is required by an e-book shop which wants to sell e-books using an e-commerce model?

9.10 Give a block diagram of an e-commerce model of an e-book shop. Give the various steps which a customer follows to buy an e-book.

9.11 Compare and contrast delivery of e-books using Internet versus a mobile communication network.

9.12 How can IPR be protected when an e-book is sold?

9.13 What infrastructure is required by an e-music shop which wants to sell music using e-commerce model?

9.14 Find out about the e-commerce model of Apple which sells music tracks and delivers it to the hand-held device called ipod.

9.15 Calculate the storage (in GB) needed to store 8000 technical books which have figures/graphs. Take a typical book to do your calculations with 8% of print area for figures. To find out the number of characters/pages, take 3 or 4 textbooks and find the average.

9.16 Find out about .jpeg compressed file format for figures. What is the compression ratio if .bmp file is compressed to .jpeg?

9.17 A library wants to digitize 5000 books in their collection to be e-delivered to its subscribers. The average size of the books is 400 pages and they are all old novels not available in word-processed form. Give the steps they should follow to create a database of books. How much storage will be needed to store the books?

9.18 A music shop wants to sell individual tracks from popular music in their shop. Each CD has 150 MP3 music tracks of average duration of 3 minutes. Find out the storage needed by each track. If the store has 5000 CDs, what is the size of database it requires to store the music?

9.19 What is the storage required to store a 2-hour movie compressed in MPEG2 and in MPEG4 (H.264)? If a Video on demand system has a database of 4000 videos, how much online disk storage will be needed to store these?

9.20 Why are videos primarily rented for viewing rather than sold? Draw a block diagram of a VOD system and explain the functions of each block.

9.21 What is the bandwidth needed to transmit MPEG2 compressed video in a VOD system? What bandwidth is required with H.264 video compression?

9.22 What are the functions of a video server in a VOD system?

9.23 Why do music/video publishers have reservations on e-distribution of audio and video?

9.24 What are the new problems which arise in protecting IPR with the emergence of computers and the Internet?

9.25 How are natural barriers to copyright protection removed in e-publishing?
9.26 What are the primary objectives of IPR law?
9.27 What rights do authors have under IPR laws?
9.28 What are the technical solutions to protect IPR?
9.29 What are the business strategies one may use to prevent IPR infringement?
9.30 What strategy did Apple use to popularize e-music on payment?
9.31 Why are business strategies better to prevent IPR infringement than using technical solutions or by amending IPR law?

OBJECTIVE QUESTIONS

Each question has four possible answers. Pick the most appropriate answer.

9.1 E-distribution of books has become feasible with the emergence of
  (a) E-book readers using e-ink technology
  (b) High speed mobile networks
  (c) The Internet
  (d) Flat screen TFT VDUs

9.2 E-distribution of multimedia has become feasible due to
  (a) The Internet
  (b) Lowering of cost of large bandwidth electronic communication
  (c) Increase in CPU power of computers
  (d) Lowering of cost of digital storage

9.3 The major difference between e-commerce of general goods and multimedia (including e-books) is
  (a) Speed of delivery
  (b) Payment method
  (c) Encryption method
  (d) Distribution is also by communication system (Internet or mobile net)

9.4 Multimedia products have unique property of being
  (a) Copyrightable   (b) Virtual
  (c) Intellectual property (d) Electronically distributable

9.5 An essential requirement of E-publishing is
  (a) Computers for word processing
  (b) The Internet
  (c) Computer networks
  (d) Electronic communication

9.6 E-commerce of e-books requires
  (i) A telecommunication distribution system
  (ii) A device for consumers to receive and read the material conveniently
  (iii) A repository of e-books
  (iv) A computer network
    (a) (i), (ii)   (b) (i), (ii), (iii)
    (c) (i), (ii), (iii), (iv)   (d) (i), (iii), (iv)

9.7 E-commerce of audio requires
   (i) A repository of MP3 tracks
   (ii) A radio
   (iii) A device to receive and hear audio
   (iv) An appropriate communication system for distribution
   (a) (i), (iii) (b) (i), (ii), (iii)
   (c) (i), (ii), (iii), (iv) (d) (i), (iii), (iv)

9.8 E-books may be produced using
   (a) Word processors (b) Scanners
   (c) Printers (d) E-book readers

9.9 For converting printed books to e-books, we need
   (i) Fast scanners
   (ii) OCR software
   (iii) Computers
   (iv) Proofreaders
   (a) (i), (ii) (b) (i), (ii), (iii)
   (c) (i), (ii), (iii), (iv) (d) (i), (iii), (iv)

9.10 In a book which has both print and several graphics, the storage needed for graphics is
   (a) Higher than that for text (b) Lower than that for text
   (c) The same as for text (d) Much lower than that for text

9.11 The advantages of converting scanned text to ASCII are
   (i) Less storage is required to store ASCII
   (ii) Contents are searchable
   (iii) It is faster to produce e-books
   (iv) It is cheaper to produce e-books
   (a) (i), (ii) (b) (i), (ii), (iii)
   (c) (i), (ii), (iii), (iv) (d) (i), (iii), (iv)

9.12 The disadvantage of converting scanned text to ASCII is that it
   (a) Is very slow
   (b) Is very expensive
   (c) Requires proofreading to correct OCR output
   (d) Requires an expensive scanner

9.13 When a 6" × 8" page is scanned with a scanner of resolution 800 × 625 bpi, the number of bytes which will be stored in bit mapped form is
   (a) 1 MB (b) 10 MB
   (c) 3 MB (d) 6 MB

9.14 When a 6" × 8" page with 80-characters per line and 50 lines is stored using ASCII, the number of bytes stored is
   (a) 10 KB (b) 4 KB
   (c) 8 KB (d) 1 MB

9.15 If a 6" × 8" page is scanned with a scanner of resolution 800 × 625 bpi, the number of bytes which will be stored if it is compressed using a compression algorithm to jpeg form is nearly
   (a) 100 KB (b) 1 MB
   (c) 500 KB (d) 200 KB

9.16 Jpeg compression of bit mapped images reduces storage required by a factor of approximately
(a) 5
(b) 15
(c) 50
(d) 100

9.17 MP3 compression of audio reduces storage required by audio tracks by a factor of approximately
(a) 2
(b) 5
(c) 15
(d) 50

9.18 A 60-minute uncompressed music CD normally requires a storage of
(a) 600 MB
(b) 6 MB
(c) 6 GB
(d) 600 KB

9.19 The number of MP3 compressed 3 minute music tracks which can be stored in 1 GB is approximately
(a) 50
(b) 500
(c) 5000
(d) 10000

9.20 A one hour uncompressed video requires approximately
(a) 100 MB
(b) 100 KB
(c) 100 GB
(d) 1000 GB

9.21 A video compressed movie using MPEG2 compression algorithm is compressed by a factor of approximately
(a) 100
(b) 300
(c) 500
(d) 1000

9.22 A video compressed movie using MPEG 4 (H.264) algorithm is compressed by a factor of approximately
(a) 100
(b) 1000
(c) 500
(d) 200

9.23 A video library digitizes 1500 movies of 2 hours duration and compresses using MPEG4 (H.264) format. It requires a disk storage of approximately
(a) 100 GB
(b) 1 GB
(c) 1000 GB
(d) 10000 GB

9.24 An e-book must be
 (i) Mobile
 (ii) Consume low power
 (iii) Easily readable
 (iv) Connected to Internet
(a) (i), (ii)
(b) (i), (ii), (iii)
(c) (i), (ii), (iii), (iv)
(d) (i), (iii), (iv)

9.25 An e-book store stocks 10000 books of average size of 300 pages and 1000 words per page. Its database size is approximately
(a) 20 GB
(b) 200 GB
(c) 2 GB
(d) 200 MB

9.26 IPR is protected in e-books by
(a) Public key encryption
(b) Plain text encryption

(c) Encryption based on e-book reader number
(d) Session key symmetric encryption

9.27 Audio may be distributed in e-commerce by
(i) Mailing a CD
(ii) Using a mobile wireless network
(iii) Using the Internet
(iv) Using a private secure network
    (a) (i), (ii)   (b) (i), (ii), (iii)
    (c) (i), (ii), (iii), (iv)   (d) (ii), (iii)

9.28 A device used by Apple to download music tracks to a customer is called
    (a) ipod   (b) iapple
    (c) imusic   (d) itrack

9.29 Video on Demand system consists of a
(i) Video storage with a server
(ii) Video distribution wide band network
(iii) Set-top box on customer's TV set
(iv) Server with information on videos available
    (a) (i), (iii)   (b) (i), (ii), (iii)
    (c) (i), (ii), (iii), (iv)   (d) (i), (iii), (iv)

9.30 Some important functions of video server of a Video on Demand system are
(i) Admission authentication and control
(ii) Encryption of video stream
(iii) Ensuring quality of service
(iv) Providing VCR functions to the customer
    (a) (i), (ii)   (b) (i), (ii), (iii)
    (c) (i), (ii), (iii), (iv)   (d) (i), (ii), (iv)

9.31 The primary functions of admission control in a VoD system are
(a) Authenticate the customer
(b) Check whether the video can be scheduled immediately for transmission without interruption
(c) Encrypting video stream
(d) Billing customer

9.32 By peer to peer file sharing, we mean
(a) Two users connected to the Internet can download files from one another if permitted
(b) Using ftp
(c) Two peers can download files from a server connected to the Internet
(d) Napster's system

9.33 Napster.com was closed down because
(a) It overloaded the Internet
(b) It distributed spam
(c) It distributed copyrighted music without payment
(d) It enabled users to infringe copyright of music producers

9.34 Intellectual Property Rights give a holder of the rights to
(i) File a case in court if his or her intellectual property is reproduced without permission

(ii) Make derivative work based on it
(iii) Protect his or her right only during his or her lifetime
(iv) Translate it to any other language
    (a) (i), (ii)   (b) (i), (ii), (iii)
    (c) (i), (ii), (iii), (iv)   (d) (i), (ii), (iv)

9.35 By intellectual property, we mean
(i) Creative work such as a book/research paper
(ii) Music recordings
(iii) Movie recordings
(iv) Works of art such as paintings
    (a) (i), (ii)   (b) (i), (ii), (iii)
    (c) (i), (ii), (iii), (iv)   (d) (i), (iii), (iv)

9.36 Music and movie producers feel threatened by the emergence of Information Technology as
(i) Music and video files are now digital bit streams and can be easily copied
(ii) Copies are as good as original
(iii) Distribution using the Internet is fast and inexpensive
(iv) Music and video files cannot be encrypted
    (a) (i), (ii)   (b) (i), (ii), (iii)
    (c) (i), (ii), (iii)   (d) (i), (ii), (iv)

9.37 Digital millennium copyright act makes it illegal to
(a) Encrypt copyrighted music/video
(b) Write programs to break encryption protection
(c) Decrypt copyrighted music/video
(d) Scramble encrypted music/video

9.38 Technical methods of copyright protection include
(i) Encryption with a key sold with product
(ii) Read only medium with copying barred
(iii) Digital watermarks
(iv) Notice prohibiting copying
    (a) (i), (ii)   (b) (i), (ii), (iii)
    (c) (i), (ii), (iii), (iv)   (d) (i), (ii), (iv)

9.39 Business solutions
(a) Are more effective than technical solutions
(b) Are less effective than technical solutions
(c) Are not different from technical solutions
(d) Are expensive to implement

9.40 Business solutions for copyright protection are
(i) License to use and not for sale
(ii) Make it cheaper to buy legal copies
(iii) Give away product free and make money on updates and service
(iv) Sell at low cost and depend on large volume
    (a) (i), (ii)   (b) (i), (ii), (iii)
    (c) (i), (ii), (iii), (iv)   (d) (i), (iii), (iv)

CHAPTER 10

Legal Framework of E-Commerce

LEARNING GOALS

In this chapter, we will learn:
1. Information Technology Act 2000.
2. About amendments to the act passed in 2008.

10.1 INFORMATION TECHNOLOGY ACT 2000

India is one of the few countries in the world which passed a new law called Information Technology Act 2000 to promote e-commerce transactions in India. The act was drafted by using the model law on e-commerce initiated by the United Nations Commission on International Trade Law (UNCITRAL).

A special law to deal with e-commerce is necessary due to several peculiar features which are specific to e-commerce world. These are:

- The Internet has no physical or national boundaries. There is no single "controller" of the Internet.
- All correspondence and documents are in electronic form. There are no handwritten or physically signed documents. There are no other identifying marks such as printed or embossed letter head, seals, thumb impressions.

- Taxation laws, international laws are not very clear between various countries.
- As the Internet is "open" there is a perceived lack of security and confidentiality unless special precautions are taken.
- Computer and communication technology is changing rapidly.
- Most common people are ignorant about technology.

The primary objectives of the Act are:

- There is a need to promote e-commerce and amend existing laws to be in tune with the new technology. Thus, there is a need to recognize electronic documents and to recognize digital signature analogous to physical signature. This will enable the conclusion of contracts and enforce rights and obligations relevant to electronic documents.
- The authenticity of digital signatures has to be regulated by an appropriate Government certifying authority to ensure their wide acceptance and enforcement of digitally signed documents in courts of law.
- To promote e-governance the act proposes Government offices and agencies to accept electronic records signed digitally.
- The act also will make consequential amendments in the Indian Penal Code, and the Indian Evidence Act 1892 which deal with offences related to documents and paper-based transactions. It proposes to amend the Reserve Bank of India Act 1934 to facilitate electronic funds transfer between the financial institutions and bankers. The Bankers' Book Evidence Act 1891 is also to be amended to give legal sanctity to bank accounts maintained in electronic form by banks.

The act has explicitly left out the following from its scope:

- Negotiable instruments
- Power of Attorney
- Trust Deed
- Will
- Contract for sale of property

The above documents are normally registered by registration offices of various states after paying appropriate stamp duties.

The major interesting aspects of the act are:

- E-mail correspondence has legal status and, therefore, it can be used as evidence in a court.
- The use of the private key and public key in encryption has been recognized as a secure method of transmitting data electronically. Consequent to this digital signature based on the private key–public key pair is recognized for signing documents.
- By signature it is assumed to be digital signature in the act. This has since been amended.
- In order to authenticate public keys a controller of public key certifying authorities has been appointed by the Government. This authority can grant licences to certifying authorities based on certain criteria. The controller's office will be repository of all public keys. The public key certificate will be issued by the certifying authorities but if required it can also be checked with the repository maintained by the controller.

- The foreign public key certifying authorities can be recognized by the controller in India.
- All applications to Government bodies can be filed in electronic form. The Government can issue licences, permits, sanctions, approvals, etc., online in electronic form. All these have to be digitally signed.
- Many archival documents which companies and government departments are required by law to be kept for a specified period can now be stored in CDROM or tapes saving precious space in buildings and enabling easy retrieval. Care must be taken to ensure that such electronically stored documents keep details which will identify the origin of the document, date and time of dispatch or receipt.

The following have been classified as offences by the Information Technology Act 2000 and punishment of imprisonment and/or fine has been specified.

- If a company's network is illegally accessed and stored data is stolen or damaged monetary claims upto Rs. 1 core can be made against the intruder. Similar claim can be made for flooding the site leading to a denial of access to a company's site.
- If a person(s) steals source code stored in a company's computer or tampers with it, a punishment of maximum 3 years imprisonment and/or Rs. 2 lakhs fine can be levied.
- If private confidential information is accessed for unlawful purposes by a person(s) it is punishable with imprisonment of up to 2 years.

The act also specifies imprisonment and fine for the following offences committed by a cyber criminal, popularly known as a hacker even though there are ethical hackers with no criminal intent who expose security flaws in systems:

- Downloading copies of or extracts of confidential data from a database without permission of the owner.
- Introducing any soft-contaminant or computer virus into any computer or a computer network.
- Altering or deleting data from a person's computer without the person's knowledge.
- Publishing obscene literature/pictures accessible from the Internet.
- Charging for services availed of by a person to another person by tampering with or manipulating accounts in a computer network.
- Using the Internet for any act which will compromise the sovereignty or integrity of India.

A legal framework has also been created for trying cyber crimes. They are:

- An adjudicating officer has been appointed to hold inquiries under the act.
- A cyber Appellate Tribunal with a high court judge as presiding officer and other experts have been appointed.
- All appeals against the order of an adjudicating officer are to be heard by the Cyber Regulations Appellate Tribunal and not by any civil court.
- All appeals against the order of the tribunal will be heard by a High Court.
- Police officers of a certain rank have been given the authority to enter any public place (for example, a cyber café) and search and arrest without warrant if they suspect that an action breaking the provisions of the IT Act is being committed. This is a draconian measure which is not probably entirely warranted.

- A Cyber Regulation Advisory Committee has been appointed to advise the controller of public key certifying authorities and the Government on matters relating to the actual implementation and working of the Act.

The Act precludes Internet Service Providers (i.e., ISPs) and other service providers from its scope.

The IT Act 2000 has several flaws which have been pointed out by several experts. They are:

- It is not clear how cyber crimes affecting computers in India committed from outside India using the Internet will be handled.
- Many provisions of the Act are such that it is not clear how they can be enforced.
- The Act does not have any provisions regarding domain names and resolving disputes on such names.
- Many cyber crimes are not defined in the Act such as cyber defamation, cyber harassment and cyber stalking.
- Privacy and protection of personal data such as medical records are not covered by the Act.
- The Act does not deal with intellectual property rights, trademarks and patents.
- The Act has no provision to punish persons or organizations sending unsolicited mail normally known as spam.

In spite of the Act having been in the statute book, very few cases have actually been investigated thoroughly and enforced due to lack of knowledge of police officers. It is being rectified by training and establishing special cyber crime police stations in several states with specially trained police officers.

10.2 INFORMATION TECHNOLOGY (AMENDMENT) ACT 2008

Based on the criticism and experience gained by using the Information Technology Act 2000, the Government set up an expert committee to review the IT Act in January 2005. The committee had representatives from the Government, IT Industry and legal experts and submitted its report in August 2005. After approval from the Government, the Information Technology (Amendment) Bill 2006 was submitted to the parliament. The parliament sent it to its standing committee which has made some more recommendations. The bill is called IT (Amendment) Bill 2008 has been debated and passed by Lok Sabha and is now an Act. Thus, what we will describe in this section is the amendments to the IT Act 2000 and how it rectifies some defects in the current law.

The main amendments are:

1. In the IT Act 2000, an e-document is affixed with a digital signature which is based on encryption with a certified public key. It is not "technology neutral". In other words, it is tied to a particular method of digital encryption. If this encryption method is found insecure by some unforeseen future technology the entire law breaks down with all the structures which has been put in place. In such a case, other methods of signing such as affixing scanned thumb print (or other unique biometric markers), using a digital watermark, etc., may have to be used. Thus, the amendment replaces the term *digital signature* by the term *electronic signature*.

This new term does not exclude digital signature as it is an electronic signature. However, the types of electronic signatures which are allowed are not specified. It will be specified from "time to time" by the Government.

2. In IT Act 2000 the controller of public key certifying authorities is a government appointee in the Department of IT who has to keep in his or her office all public key certificates to allow anyone to access the database (or repository) of public key certificates to authenticate the certificate. This provision has been amended allowing the certifying authorities to provide public key certificates. The purpose is to relieve the controller's office whose public key database could become huge. However, authenticity of a certificate given by a responsible government official has better credibility in public's perception.

3. The other set of important amendments relate to the protection of the privacy of personal data. Interestingly no specific law in the statute books currently in India addresses directly this issue of data privacy. However, the Supreme Court in a recent ruling has opined that privacy is a right flowing from the constitutional guarantee of right to life. The 2008 Act places responsibility of ensuring security of personal data of individuals handled by a company. If through their negligence in securing the data of individuals "sensitive private data" is accessed by unauthorized persons, the company is liable to pay a compensation of up to Rs. 5 crores to the affected individual. It does not define what constitutes "sensitive private data".

4. The amended Act also penalizes service providers who collect personal data to provide, for example, a free service, from disclosing it to anyone else with intent to cause injury to the individual.

5. Another amendment relates to circulation of indecent pictures or videos of individuals (e.g., nude pictures) without their permission. Anyone doing this can be jailed for an year and fined up to Rs. 25 lakhs.

6. The term hacking which was used in IT Act has been replaced by the more accurate term *computer related offence*. As we pointed out while discussing IT Act 2000, the term *hacker* does not necessarily mean a cyber criminal.

7. The provision of the IT Act 2000 which relates to publishing and transmission of pornography has been substantially changed. Only entities which *intentionally or knowingly* are involved in publishing or transmission of pornographic material are punishable. Intermediaries have been excluded from the ambit of this law. For example, a search engine company like Google is not liable for pornographic material which is retrieved using their tool. However, if active collusion of the intermediary is proved in publishing and transmission of pornographic material, then it is liable. If an intermediary is informed about such an objectionable material, it must remove the material.

8. If an offence is committed by a company, the person managing its affairs (such as CEO) is normally liable. The amendment modifies this clause. A manager/director is punishable only if it is proved that the person connived in committing the offence and failed to prevent it.

9. The IT Act 2000 gave arbitrary powers to the police. Under its provision a police officer can enter, search and arrest an individual from a public place if he or she suspects that the IT Act is being violated. This provision has now been removed.

10. The IT Act 2000 provides an appellate authority to appeal against the rulings of the controller. It had one person appointed by the Government. It has been changed to "Cyber Appellate Tribunal" which would consist of a chairperson and other members to be appointed by the Government. One member of the tribunal will be a judicial member. The Government will also appoint an examiner (or examiner's office) to give expert opinion on electronic form of evidence.

The following material has been based on an analysis of the proposed bill which has been published by PRS Legislative Research, New Delhi (www.prsindia.org.http://prsindia.org/legis-page.php?bill-id=93). It is an analysis of some of the issues not properly addressed by the 2006 bill but have since been addressed in the 2008 Act.

- Currently telephones can be tapped and regular mail intercepted only to protect sovereignty of the country, national interest, etc. The 2008 amendment allows e-mail communications also to be intercepted by the police for investigation of *any offence*.
- As pointed out earlier there is no current law on privacy in India except under the constitutional guarantee of right to life. The IT Act 2000 does not specify what personal information may be collected, how it can be processed, used and disseminated. The 2008 Act provides compensation to persons whose personal data has been used unlawfully without permission of the individual. It does not, however, address the issue of breach of privacy.
- Copying and destroying personal data without permission of the individual is punishable under the amended act. If an employee of a company who is authorized to access personal data misuses it, there is no provision to deal with it.
- The 2008 Act defines child pornography. Using computers/communication device to propagate child pornography will attract exemplary punishment.
- The Act adds to the definition of intermediary telecom networks, Internet and web hosting service providers, search engines, online payment and auction sites. It defines cyber café as any facility which provides Internet access to the general public as part of their business and includes them as an intermediary.
- New offences have been added as part of the amended act. They are: sending offensive messages using a computer or a mobile phone, disclosing information in breach of a lawful contract, cheating by using a computer, sending nude pictures of persons without their permission.

Other provisions which are now added are:

- Punishment has been prescribed for receiving stolen communication devices including computers.
- Dishonestly using electronic signature and password is an offence and it is punishable.
- If a person cheats by impersonating using a communication device or a computer resource, it is punishable.
- Cyber terrorism has been defined and exemplary punishment has been prescribed for terrorist acts using computers or communication device.
- The amended act provides punishment for sending spam (i.e., unsolicited e-mail on a mass scale).

- Appropriate officials have been empowered to monitor and collect traffic data to ensure cyber security including virus and other computer contaminant distribution.
- Appropriate officials have been empowered to issue directions to ISP etc., to block public access of any information dissemination through a computer communication device which in their considered opinion is detrimental to the sovereignty, integrity and defence of India. Blocking of information can also be ordered to prevent incitement to the commission of cognizable offence related to security, sovereignity, etc., of India.
- An Indian computer emergency response team is to be formed to collect, analyze and disseminate information on cyber incidents and to forecast, alert and take emergency measures to handle such incidents. It will also issue guidelines, advisories, vulnerability notes to all concerned from time to time.

The standing committee of the parliament which studied the bill made some suggestions which are listed below:

- Punishment for cyber crimes committed outside India cannot be enforced. It suggests that India should take the initiative to convene an International convention on the issue of cross border cyber crimes.
- Due diligence obligations must be enforced on intermediaries who deal with personal data before giving them immunity particularly in areas such as online auction sites and online market places.
- The complicated adjudication process proposed in the bill for obtaining compensation for various crimes should be simplified.
- The jurisdiction of the appellate tribunal and civil courts in various cases should be clarified.
- The government along with industry should initiate training programs for all the entities dealing with cyber crimes.

SUMMARY

1. The IT Act was passed in 2000 to promote e-commerce.
2. The Act made digital signature using asymmetric encryption legal for signing e-documents.
3. E-mail correspondence was given legal status.
4. To support digital signature a controller of public key certifying authorities was appointed to be a repository of public keys of organizations issued by certifying authorities.
5. E-documents filed with the Government and those stored in CD-ROM etc., were given legal status.
6. Several offences were defined in the IT Act 2000 which include illegally accessing computers of entities and damaging or stealing data/source code, creating and distributing viruses, publishing obscene material, using the Internet to compromise the sovereignty or integrity of India.
7. A legal framework was created to try cyber crimes.

8. Certain powers were given to law enforcement authorities to prosecute and/or prevent cyber crimes.
9. There were several flaws in the Act which led to the Government in 2005 to prepare an IT Act amendment bill and submit it to the parliament. It was examined by the standing committee of the parliament whose report was submitted in 2006. The Act has been amended in 2008.
10. The major amendments in the amended act are to replace the terms *digital signature* by *electronic signature* to make the Act technology neutral, penalize companies which are negligent in handling personal data and to protect intermediaries from liabilities for third party data/content made available to them.
11. The 2008 Act provides punishment for child pornography, cyber terrorism, distributing spam, distributing nude photographs without permission, stealing passwords, receiving stolen computers, etc.
12. It empowers appropriate authorities to read e-mails and block web sites under certain circumstances.
13. Several of the problems of the IT Act 2000 are still not addressed including protecting privacy, domain name disputes and cyber crimes committed outside India.

EXERCISES

10.1 Why is a special law needed to deal with e-commerce?
10.2 What are the primary objectives of the IT Act 2000?
10.3 What are some legal documents which are excluded by the IT Act 2000?
10.4 Under the IT Act 2000 does e-mail have a legal status?
10.5 Does the law require legal e-documents to be signed? What is the signature which is recognized in the IT Act 2000?
10.6 Who certifies the public keys of organizations/individuals?
10.7 What is the role of the controller of public key certifying authorities? Who appoints the controller?
10.8 Are foreign public key certifying authorities recognized in India? If so, by whom?
10.9 Can applications to the Government of India be filed in electronic form?
10.10 Should archival documents which are required by law to be preserved for a certain period of time be only paper documents? If not, in what other forms can they be kept? Are there any requirements to be met in these forms?
10.11 List cyber crimes which have been identified and are punishable under the IT Act 2000.
10.12 Is there any punishment for those who create and distribute viruses?
10.13 Is there any punishment for those who create and distribute spam?
10.14 Which is the authority which has been empowered to enquire into cyber crimes and punish them? What is the composition of such an authority?
10.15 To whom can one appeal against punishment awarded for cyber crimes? What is the composition of such an authority?

10.16 If cyber crimes are committed by facilities provided by intermediaries such as ISPs are the intermediaries punishable? If they are publishable under what circumstances can they be punished?

10.17 Can domain name disputes be settled under the IT Act 2000? If yes, who is the authority?

10.18 Is private data of individuals protected by the IT Act 2000? If yes, how is the breach of privacy punished?

10.19 Why have amendments been proposed to the IT Act 2000?

10.20 Is digital signature essential to authenticate e-documents in the amended 2008 Act? If not, what is the new type of signature? Why has this amendment been proposed?

10.21 In the amended Act which organization has to keep a repository of public key certificates? Why has this amendment been proposed? Discuss the pros and cons of this amendment.

10.22 How is privacy of personal data ensured in the amendment Act? Does it define private data?

10.23 How are intermediaries treated in the amended Act regarding transmission of pornographic data?

10.24 What powers do police have under the IT Act 2000 to apprehend cyber crime? Has it been proposed to be amended? If yes, what is the amendment?

10.25 Under what circumstances can e-mail correspondence be intercepted by law enforcement agency as per the amended Act?

10.26 What is the definition of cyber café in the amended Act?

10.27 Is there any provision to deal with cyber terrorism in the IT Act 2000? Has this been amended?

10.28 Is there any cyber law to deal with spam in the amended Act?

OBJECTIVE QUESTIONS

Each question has four possible answers. Pick the most appropriate answer.

10.1 The IT Act 2000 was enacted to give legal status to
   (a) The Internet   (b) E-commerce
   (c) World Wide Web   (d) Credit card transactions

10.2 The IT Act 2000 gives legal status to
   (a) Digital signature   (b) Electronic signature
   (c) Biometric signature (d) Scanned handwritten signature

10.3 E-documents are legal as per the IT Act 2000 if they are
   (a) Encrypted with a private key   (b) Encrypted with a public key
   (c) Digitally signed   (d) Electronically signed

10.4 The authenticity of digital signature may be established by referring to
   (a) Authenticating authority
   (b) Controller of public key certifying authorities
   (c) Public key certifying authorities
   (d) Cyber adjudicating officer

10.5 E-mail correspondence has legal status
   (a) Only if digitally signed
   (b) Only if encrypted with the public key
   (c) By the very act of sending it to an addressee
   (d) Only if encrypted with the private key

10.6 The IT Act 2000 has left out the following from its scope:
   (i) Power of attorney
   (ii) Contract for sale of property
   (iii) Will
   (iv) Maintaining archival documents
      (a) (i), (ii)   (b) (i), (ii), (iii), (iv)
      (c) (i), (iii), (iv)   (d) (i), (ii), (iii)

10.7 The IT Act 2000 has left out of its scope
   (i) Spam
   (ii) Trust deed
   (iii) Domain name disputes
   (iv) Distributing viruses
      (a) (i), (ii)   (b) (ii), (iii)
      (c) (i), (ii), (iv)   (d) (i), (ii), (iii)

10.8 The IT Act 2000 recognized the legality of
   (i) Foreign public key certifying authority
   (ii) Archival documents stored in CD-ROM
   (iii) Digitally signed e-documents submitted to Government bodies
   (iv) Will written in a CD-ROM
      (a) (i), (ii), (iii)   (b) (i), (ii), (iv)
      (c) (i), (ii), (iii), (iv)   (d) (i), (iii), (iv)

10.9 The following are offences under the IT Act:
   (i) Breaking into a company's web site and mutilating it
   (ii) Downloading source code from a company's site without their permission
   (iii) Downloading catalogues of goods advertised by a company for sale
   (iv) Charging for services availed by a person to another person
      (a) (i), (ii)   (b) (i), (ii), (iii)
      (c) (i), (ii), (iii), (iv)   (d) (i), (ii), (iv)

10.10 Police officers of a certain rank have the power to arrest a person under IT Act 2000
   (a) With a warrant from a magistrate if the person commits a cyber crime
   (b) If he or she is asked to do it by a cyber adjudicating officer
   (c) If a person sends pornographic material from his or her home computer
   (d) Without warrant if an action breaking the IT Act is suspected to be committed by the person in a cyber cafe

10.11 The IT Act 2000 is silent on
   (i) Cyber crimes affecting computers in India from outside India
   (ii) Sending pornographic material using the Internet
   (iii) Cyber defamation

(iv) Protecting personal private data gathered by organization
    (a) (i), (ii)   (b) (i), (ii), (iii)
    (c) (i), (iii), (iv)   (d) (i), (ii), (iii), (iv)

10.12 The main amendments proposed in the IT Act amendment 2008 are
  (i) To make digital signature not relevant
  (ii) To include along with digital signature any electronic signature
  (iii) To relieve the controller of public key certifying authorities from maintaining a repository of public keys
  (iv) Punishment for circulating indecent pictures of persons using the Internet without his or her permission
    (a) (i), (ii)   (b) (i), (ii), (iii)
    (c) (ii), (iii), (iv)   (d) (i), (ii), (iii), (iv)

10.13 If a company commits a cyber crime, its
  (a) Directors are liable
  (b) CEO is liable
  (c) Owners are liable
  (d) CEO is liable only if the crime is committed with his or her knowledge

10.14 A service provider is punishable if the provider circulates personal data
  (a) Without the permission of the owner
  (b) Only if it is for unlawful purposes
  (c) Only if it is indecent
  (d) Only if it is defamatory

10.15 If a company does not secure its database with personal data, it is punishable if the personal data is
  (a) Sensitive   (b) Indecent
  (c) Confidential   (d) Any of the above

10.16 E-mail correspondence can be intercepted by the police
  (a) Only to protect the sovereignty of the country
  (b) Only to protect integrity of the country
  (c) Investigating any cognizable offence
  (d) Only with a warrant from a magistrate

10.17 For certain cyber offences intermediaries are not liable. The IT Act amendment 2008 defines intermediaries as
  (i) Telecom network providers
  (ii) Web hosting services
  (iii) Search engine developers
  (iv) Cyber cafes
    (a) (i), (ii)   (b) (i), (ii), (iii)
    (c) (i), (ii), (iii), (iv)   (d) (i), (iii), (iv)

10.18 The IT Act 2008 addresses
  (i) Terrorism
  (ii) Child pornography
  (iii) Spam
  (iv) Circulating/Hosting pornography
    (a) (i), (ii), (iii)   (b) (i), (ii)
    (c) (i), (ii), (iii), (iv)   (d) (i), (iii), (iv)

References

1. Kalakotta, R., and Whinston, A.B., *Frontiers of Electronic Commerce*, Addison-Wesley, 1999.
2. Turban, E., Lee, J., King, D. and Chung, H.M., *Electronic Commerce: A Managerial Perspective*, Pearson Education Asia, 2001.
3. Awad, E.M., *Electronic Commerce—From Vision to fulfillment*, 3rd edition, PHI Learning, 2007.
4. Joseph, P.T., *E-commerce: An Indian Perspective*, 3rd edition, PHI Learning, 2008.
5. Rajaraman, V., *Building blocks of e-commerce*, Sadhana, Vol. 30, Parts 2, 3, April–June 2005, pp. 89–117.
6. Rajaraman, V., *Fundamentals of Computers*, 4th edition, PHI Learning, 2008.
7. Comer, D.E., *The Internet Book*, 3rd edition, PHI Learning, 2001.
8. Stamper, D.A., and Case, T.L., *Business Data Communications*, 6th edition, Pearson Education Asia, 2004.
9. Sebesta, R.W., *Programming the World Wide Web*, Pearson Education Asia, 2004.
10. Landau, S., *Designing Cryptography for the New Century*, Communications of ACM, Vol. 35, No. 5, pp. 115–128, 2000.
11. Stallings, W., *Cryptography and Network Security*, 3rd edition, Pearson Education Asia, 2003.
12. Kalakotta, R., and Robinson, M., *E-Business: Roadmap for Success*, Addison-Wesley 2000.
13. Amor, D., *The E-Business Revolution*, Pearson Education Asia, 2000.

14. Chaum, D., *Achieving Electronic Privacy, Scientific American*, pp. 96–101, 1992 (Describes Chaum's blinding protocol).
15. Minoli, E.M., and Minoli, D.M., *Web Commerce Technology Handbook*, Tata McGraw-Hill, 1999.
16. Lynch, D.C., Lundquist, L., *Digital Money: The new era of Internet Commerce*, John Wiley, 1996.
17. Marchal, B., *XML by Example*, PHI Learning, 2001.
18. Maruyama, A., Tamura, K., Uramuto, N., *XML and Java*, Addison-Wesley, 2000.
19. Pardi, W.J., *XML in Action*, Microsoft Press, 1999.
20. Goldfarb, C.F., and Prescot, P., *The XML Handbook*, 3rd edition, Pearson Education Asia, 2001.
21. Varshney, V., Vetter, R.J., Kalakotta, R., *Mobile Commerce: A new Frontier*, IEEE Computer, Vol. 35, No. 10, pp. 32-38, 2000. (This issue also has more papers on M-Commerce.)
22. Singal, S., *et al.*, *The Wireless Application Protocol*, Addison-Wesley, 2001.
23. Rajaraman, V., *Analysis and Design of Information Systems*, 2nd edition, PHI Learning, 2008.
24. Denning, D.E., *Information Warfare and Security*, 2nd edition, Addison-Wesley, 1999.
25. Electronic Money, IEEE Spectrum (Special Issue), Vol. 34, No. 2, February 1997.
26. *Money: What it is and how it works*, http://wfhammel.cnchost.com/seignorage.html
27. Samuelson, P., *Why the Anti-circumvention Regulation need Revision*, Communications of ACM, Vol. 42, No. 9, pp. 17–21, 1999.
28. Paulson, L.D., *Copyright Ruling Generates Concern*, IEEE Computer, Vol. 34, No. 1, p. 30, 2001.
29. Duggal, P., *Cyberlaw in India – an Analysis*, Sakshar Publications, 2000.
30. McGrath, S., *XML by Example: Building E-commerce Applications*, Prentice Hall Inc., 1998.
31. Rajaraman, V., *Introduction to Information Technology*, PHI Learning, 2007.
32. Steinmetz, R., and Nahrstedt, K., *Multimedia: Computing, Communications and Applications*, Prentice Hall Inc., 1995.
33. Kugle, L., *Ubiquitous Video*, Communications of ACM, Vol. 51, No. 9, pp. 14–15, 2008.

Answers to Objective Questions

CHAPTER 1

1.1 (d)	1.2 (c)	1.3 (a)	1.4 (a)	1.5 (c)
1.6 (a)	1.7 (a)	1.8 (c)	1.9 (d)	1.10 (c)

CHAPTER 2

2.1 (c)	2.2 (b)	2.3 (a)	2.4 (a)	2.5 (d)
2.6 (a)	2.7 (b)	2.8 (b)	2.9 (c)	2.10 (a)
2.11 (b)	2.12 (b)	2.13 (d)	2.14 (d)	2.15 (b)
2.16 (a)	2.17 (b)	2.18 (d)	2.19 (c)	2.20 (d)
2.21 (a)	2.22 (c)	2.23 (b)	2.24 (c)	2.25 (b)
2.26 (d)	2.27 (a)	2.28 (b)	2.29 (a)	2.30 (a)
2.31 (d)	2.32 (b)			

CHAPTER 3

3.1 (d)	3.2 (b)	3.3 (c)	3.4 (b)	3.5 (a)
3.6 (b)	3.7 (d)	3.8 (a)	3.9 (a)	3.10 (c)

3.11 (c)	3.12 (c)	3.13 (c)	3.14 (b)	3.15 (c)
3.16 (c)	3.17 (d)	3.18 (b)	3.19 (a)	3.20 (b)
3.21 (c)	3.22 (a)	3.23 (c)	3.24 (a)	3.25 (b)
3.26 (b)	3.27 (c)	3.28 (a)	3.29 (c)	3.30 (d)
3.31 (c)	3.32 (d)	3.33 (a)		

CHAPTER 4

4.1 (c)	4.2 (d)	4.3 (b)	4.4 (b)	4.5 (c)
4.6 (b)	4.7 (d)	4.8 (a)	4.9 (d)	4.10 (b)
4.11 (a)	4.12 (c)	4.13 (c)	4.14 (b)	4.15 (c)
4.16 (b)	4.17 (c)	4.18 (b)	4.19 (a)	4.20 (b)
4.21 (d)	4.22 (c)	4.23 (b)	4.24 (a)	4.25 (c)
4.26 (c)	4.27 (a)	4.28 (d)	4.29 (a)	4.30 (b)
4.31 (b)	4.32 (b)	4.33 (a)	4.34 (c)	4.35 (b)
4.36 (a)				

CHAPTER 5

5.1 (c)	5.2 (d)	5.3 (c)	5.4 (c)	5.5 (b)
5.6 (a)	5.7 (a)	5.8 (b)	5.9 (a)	5.10 (c)
5.11 (b)	5.12 (d)	5.13 (b)	5.14 (a)	5.15 (c)
5.16 (d)	5.17 (a)	5.18 (b)	5.19 (b)	5.20 (b)
5.21 (c)	5.22 (a)	5.23 (b)	5.24 (d)	5.25 (c)
5.26 (d)	5.27 (a)	5.28 (c)	5.29 (c)	5.30 (a)
5.31 (d)	5.32 (b)	5.33 (c)	5.34 (b)	5.35 (a)
5.36 (c)	5.37 (c)	5.38 (c)	5.39 (d)	5.40 (c)
5.41 (b)	5.42 (b)	5.43 (a)	5.44 (b)	5.45 (d)
5.46 (c)	5.47 (c)	5.48 (d)	5.49 (b)	

CHAPTER 6

6.1 (a)	6.2 (c)	6.3 (d)	6.4 (a)	6.5 (b)
6.6 (d)	6.7 (c)	6.8 (b)	6.9 (c)	6.10 (b)
6.11 (a)	6.12 (b)	6.13 (b)	6.14 (a)	6.15 (c)
6.16 (a)	6.17 (a)	6.18 (b)	6.19 (c)	6.20 (c)
6.21 (a)	6.22 (c)	6.23 (a)	6.24 (c)	6.25 (b)
6.26 (c)	6.27 (b)	6.28 (d)	6.29 (d)	6.30 (a)

6.31 (c)	6.32 (c)	6.33 (b)	6.34 (d)	6.35 (d)
6.36 (c)	6.37 (d)	6.38 (d)	6.39 (b)	6.40 (b)
6.41 (c)	6.42 (a)	6.43 (c)	6.44 (d)	6.45 (a)
6.46 (a)	6.47 (b)	6.48 (a)	6.49 (b)	6.50 (a)
6.51 (b)	6.52 (d)	6.53 (d)	6.54 (b)	

CHAPTER 7

7.1 (c)	7.2 (d)	7.3 (b)	7.4 (a)	7.5 (d)
7.6 (d)	7.7 (a)	7.8 (a)	7.9 (d)	7.10 (c)
7.11 (d)	7.12 (d)	7.13 (b)	7.14 (a)	7.15 (c)
7.16 (b)	7.17 (b)	7.18 (a)	7.19 (a)	7.20 (b)
7.21 (c)	7.22 (b)	7.23 (a)	7.24 (c)	7.25 (d)
7.26 (c)	7.27 (a)	7.28 (b)	7.29 (b)	7.30 (d)
7.31 (a)	7.32 (b)			

CHAPTER 8

8.1 (d)	8.2 (c)	8.3 (b)	8.4 (a)	8.5 (c)
8.6 (b)	8.7 (a)	8.8 (b)	8.9 (c)	8.10 (a)
8.11 (b)	8.12 (b)	8.13 (d)	8.14 (b)	8.15 (a)
8.16 (d)	8.17 (c)	8.18 (c)	8.19 (b)	8.20 (b)
8.21 (c)	8.22 (b)	8.23 (d)	8.24 (a)	8.25 (d)
8.26 (c)	8.27 (b)	8.28 (d)	8.29 (c)	8.30 (d)
8.31 (a)	8.32 (b)	8.33 (c)	8.34 (d)	8.35 (c)
8.36 (a)	8.37 (d)	8.38 (b)	8.39 (c)	8.40 (c)
8.41 (b)	8.42 (a)	8.43 (d)	8.44 (a)	8.45 (b)
8.46 (b)	8.47 (c)	8.48 (d)		

CHAPTER 9

9.1 (a)	9.2 (b)	9.3 (d)	9.4 (d)	9.5 (b)
9.6 (b)	9.7 (d)	9.8 (a)	9.9 (c)	9.10 (a)
9.11 (a)	9.12 (c)	9.13 (c)	9.14 (b)	9.15 (d)
9.16 (b)	9.17 (c)	9.18 (a)	9.19 (b)	9.20 (c)
9.21 (a)	9.22 (d)	9.23 (c)	9.24 (c)	9.25 (a)
9.26 (c)	9.27 (d)	9.28 (a)	9.29 (c)	9.30 (c)

9.31 (b)   9.32 (a)   9.33 (d)   9.34 (d)   9.35 (c)
9.36 (b)   9.37 (b)   9.38 (b)   9.39 (a)   9.40 (c)

CHAPTER 10

10.1 (b)   10.2 (a)   10.3 (c)   10.4 (b)   10.5 (a)
10.6 (d)   10.7 (b)   10.8 (a)   10.9 (d)   10.10 (d)
10.11 (c)   10.12 (c)   10.13 (d)   10.14 (b)   10.15 (a)
10.16 (c)   10.17 (c)   10.18 (c)

Index

10 Base T, 21
100 Base T, 23

Active documents, 74
ADSL modem, 25
Advanced Encryption Standard (AES), 91
Automated cheque clearance, 125
Automated clearing house (ACH), 125

Backbone Hub, 22
Bridge, 23
Broadband connection, 25
Browser, 58
   multiple windows, 72

Cable network, 26
Cascading style sheets, 74
CDMA, 177
CDMA 1xEVDO, 180
Common Gateway Interface (CGI), 73

Communication protocol, 19
Cookies, 73
Credit card payment, 118
   method for manual system, 118
   parties involved in, 118
   secure electronic transaction protocol, 121
   using secure socket layer, 120
Cryptography, 85
CSMA/CD, 20
Cyber crimes
   appellate tribunal, 231
   committed outside India, 232
   trial of, 228–229
Cybersquatting, 48
Cypher text, 84

DES encryption algorithm, 88–89
Diffie–Hellman public key encryption, 96
Digital Encryption Standard (DES), 87
Digital signature, 101
Digitizing audio, 210
Digitizing books, 209

Digitizing video, 210
Distribution of
　audio, 212
　e-books, 209
　video, 213, 215
Document Type Definition (DTD), 157, 159
Domain name, 43
Dual signature scheme, 122

E-books, 209
　distribution, 211
E-commerce
　advantages of, 10
　B2B, 3, 6
　B2C, 3
　C2C, 3, 6
　definition of, 2
　disadvantages of, 11
　G2B, 3
　G2C, 3
　infrastructure for, 16
　of multimedia, 207
　types of
E-commerce layered architecture, 17
E-commerce of multimedia, 206
EDI formatted data
　gateway programming service, 156
　transport of, 155
　using VAN, 155
EDI standard, 153
　ANSIX12, 153
　EDIFACT, 153
Electronic cash, 137
　anonymous, 134
　issue, 132
　smart card based, 135
　spending, 133
Electronic cheque format, 131
Electronic cheque payment, 127
Electronic clearance of pay order, 129
Electronic clearing service (ECS), 126
Electronic data interchange (EDI), 152
Electronic funds transfer (EFT), 124
Electronic signature, 229
E-mail, 68, 70
Encryption, 85
E-payment, 116
　comparison of methods, 139
　for information goods, 138
　system requirements, 117
E-publishing, 208

Ethernet, 19
Ethernet switches, 23
Extensible Markup Language (XML), 157
Extranet, 18, 46

File transfer protocol (ftp), 68, 69
Firewall, 18, 47
Full duplex communication, 23

General Packet Radio Service (GPRS), 180
Gigabit Ethernet, 21
GPRS, 180
GSM, 177

Hash function, 102
　MD5, 102
　SHA-1, 103
Home page, 162–163
Hub, 20–21
Hypertext, 58
Hypertext marked language (html), 63–64
Hypertext transfer protocol (http), 58, 60

ICANN, 44
IEEE802.11b, 28
i-mode, 185
Information Technology (Amendment) Act 2008, 229
　main amendments, 229–231
Information Technology ACT 2000, 226
　aspects of, 227
　flaws in, 229
　objectives of, 227
　offences under, 228
Intellectual property rights (IPR), 215
Internet, 18, 40
Internet protocol, 40, 42
Internet Service Provider (ISP), 24
Intranet, 18, 45
IP address, 42
IP packet, 41
IPSEC, 46
IPv4, 50
IPv6, 50
ISDN service, 26

Local Area Network (LAN), 19

Markup languages, 63
M-commerce, 173
 applications of, 175
 layered architecture, 175
 location dependent, 176
Micro-payments, 130
Microwave network, 28
Mobile banking, 193
 WAP gateway based, 145
Mobile communication, 177
 arhcitecture of, 178
 base station, 177–178
 CDMA, 177
 establishing connections in, 179
 GSM cellular system, 177
 hexagonal cells, 178
 Mobile Telecommunication Switching Office (MTSO), 178
Mobile payment methods, 190
 SIM card based, 190
 SMS based, 190
 with WAP enabled device, 191
Modem, 24

Net Bill, 138
Network Interface Card (NIC), 19

Packet switching, 40
Pay Pal, 137
Payment gateway, 137
Permutuation, 85–86
Plain text, 84
Private communication networks, 31
Protecting IPR
 business solutions, 217–218
 technical solutions, 21, 67
Public key certifying authority, 99
Public key encryption, 92
Public Switched Telephone Network (PSTN), 24

Router, 23–24
RSA and DES, 98
RSA encryption, 93

Satellite communication, 25, 30
Search engine, 62

Secret key, 85
Secure electronic transaction protocol (SET), 121
Secure e-mail, 103
 S/MME, 103
Secure hypertext protocol (https), 104
Secure socket layer, 104
Secure wireless connectivity, 188
Set top box, 214
Short Messaging Service (SMS), 181
Smart card payment, 135
SMS, 181
Splitter, 25
Standard Generalised Markup Language (SGML), 63
Storage requirements of
 audio, 210
 e-books, 209
 video, 210
Structured electronic documents, 151, 157
Substitution, 85
Supply chain management, 12
Symmetric data encryption, 84
Symmetric key encryption, 84
 3DES, 90
 DES, 88, 89
 IDEA, 103

TCP/IP, 18, 43
Telnet, 68–69
Transmission Control Protocol (TCP), 43
Triple DES encryption, 90

Universal Resource Locator (URL), 60

Value Added Networks (VAN), 154
 functions of, 155
Video-on-demand, 213
 components of, 213
Virtual Private Network (VPN), 18, 46
VSAT, 31

WAP, 176, 181
 comparison with IP, 183
 gateway, 181, 183, 189
 protocol stack, 181–182
WAP proxy, 184

246 Index

WAP stack
  parts of, 182
  WAE, 182
  WSP, 182
  WTLS, 182
  WTS, 182
Web crawler, 63
Web page, 64
  with table, 66
Web page design, 162
Website, 162
WiFi, 28, 175
Wireless Application Protocol (WAP), 176, 181
Wireless hotspot, 27

Wireless Markup Language (WML), 184, 185
Wireless networks, 27
WML, 185
  card, 185–186
  decks, 185
  display, 187
World Wide Web, 58

XHTML, 187
XLL, 157–158
XML, 157
XML and HTML, 160
XSL, 157–158, 161